Tales from a City Farmyard

by

Patrick Boland

Published in 1995 by
Patrick Boland,
Glenmalure,
Lansdowne Village,
Dublin 4, Ireland.

Out of respect for the privacy of the individuals concerned, some names in the book have been changed.

British Library Cataloguing in Publication Data
A catalogue record for this book is available from the British Library.

ISBN 0-9526737-0-3 Paperback

10 9 8 7 6 5 4 3 2

Cover illustration by Ken Drakeford

Cover design and colour separations by Typeform Ltd.

Printed by Colour Books Limited,
Baldoyle Industrial Estate,
Dublin 13, Ireland.

Contents

DEDICATION

I would like to dedicate this book to my late mother and father, with happy, sometimes sad, but at all times loving memories of them, and the magical times we were fortunate enough to have lived through, and to have shared.

To my Mother and Father

— somewhere over the rainbow —

Ma:
I got the briefcase and the cheque book.
I am afraid though, the gypsy's fortune-
telling was wide of the mark — so far.
But I hope you would have been proud
of me nevertheless.

Da:
I kept the flag flying. It wasn't easy,
for the world is not as you thought it was.
And although I looked everywhere, your
standards were nowhere to be found. But
you were there, everywhere, all the time.

A SPECIAL THANK YOU

Were it not for the practical support and encouragement of a special group of friends, as well as the whole-hearted co-operation of a small but talented group of professional people in publishing and related businesses, this book would not have been produced. Indeed, were it not for the urgings of two of my friends in particular, it would not even have been written.

To set out the many ways in which these friends and colleagues helped me would make quite a big chapter in itself. But knowing how unassuming they all are, it is not a chapter they would like me to write, nor indeed would they thank me for doing so. So, very simply, but most sincerely, I would like to say thank you to:

Phelim O'Doherty, Tom Richards, Anne O'Neill, Bríd Ingoldsby, Paddy Burns, Ken Drakeford, Teenie Moloney, Seán Roche, Gillian Flood, Hughie Richardson, Arnold Roberts, Bernadette Smyth, Anil Jethwa and Kevin Massey.

I would also like to extend my thanks to the many others, too numerous to mention, who helped in different ways.

THE TYRONE GUTHRIE CENTRE, AT ANNAGHMAKERRIG

The Tyrone Guthrie Centre is, to me,

… the inspirational walks through the forest and by lake shore;

… the after-dinner conversations with fellow travellers who generously shared their experiences and offered encouragement;

… the unobtrusive, but essential, support of Bernard, Mary, Doreen, Ann and Ingrid, given so warmly;

… the solitude, when the heart and mind settle and the creative juices flow;

… above all, it is the solitude.

Anaghmakerrig is where I finished this book.

Thank you, Tyrone Guthrie, wherever you are.

THE FARMYARD

The only unusual things about the farmyard which joined onto the house in which I was born are that it was in The Liberties, the oldest part of Dublin, and there was no farm to go with it.

The farmyard from which my father earned a living was large enough to hold about one hundred pigs, six horses, forty hens, and one very happy cock named Errol, who strutted around with a perpetual smile on his face, and was always too tired or too busy to crow at dawn. Like his namesake Errol Flynn, he had bags under his eyes, and a lively interest in "chicks," albeit of a different kingdom. There were also three dogs, one duck and umpteen cats.

A brief appearance was made by a cow to satisfy my mother's short-lived desire for milk "straight from the source," and by a goat, named Hopalong after a cowboy of the era. The goat, given to my father in settlement of a business debt, was to stretch his emotions from joy to anger, but that's a story of its own. On the other hand, a cat named Albert brought only anger and frustration into my father's life, and we all knew that, despite Dad's love of animals, he would not have shed a tear on the departure of Albert, however it could be arranged. There were also furry little animals that nobody loved ... and that sometimes got into the house.

I was born in the Thirties, and from the age of two was reared "on the street". The street in question was Cork Street, where, I was told, some houses dated as far back as the early eighteenth century. Whereas our yard, to the best of my knowledge, was the only one with such a large number of varied animals, there were also two dairy yards and a pig yard.

The street was where it all, or most of it, happened for me. Traffic was never a problem — the occasional vehicle, usually horse-drawn, could be heard coming a long way off. My recollections are mostly of summertime, when I could play from early morning when I rushed out to the street clean and shiny, until late at night when I was dragged in to the house, filthy dirty, and sometimes black and blue from whatever the situations of play might have thrown up.

Meal times meant nothing, only that they were an interruption to our games. When my mother would come to the hall door to call me in for dinner, I would try to hide. When after much pleading and argument I was eventually got to the dinner table, the ceremony of eating the meal took no more than three minutes, and then I was free again to go back out on the street.

Apart from the street, the other great place of freedom, fun and adventure was our farmyard. In the middle of the farmyard there was a huge round wooden water barrel where we sailed our home-made boats, and jumped into in our bathing togs on hot summer days. We would sit in the hayshed and "tell pictures," or hide in a million secret places where Mammies or Daddies, and maybe others with ulterior motives, could not find us.

It was a great source of entertainment for me and my pals, for apart from giving access to the animals to play with, it also provided great hiding places, places to play games, and a harness room and hayshed in which to relax and tell stories. The harness room, where obviously the harnesses were kept, also held barrels and bags of oats, bran, cleaning materials and containers of liquids and powders for all kinds of purposes. The room had a fireplace where my pals and I could light a big fire to stay warm on cold winter evenings, and toast bread, fry sausages if we had any, and brew up tea in the billy cans that we had for the drivers. In addition to that, there was a lovely smell from the leather harness and the polishes and oils which the Da used on them. Another pleasant smell was that which came from Mr Mullery's bakery which was directly next door to the harness room. On cold winter evenings, sitting comfortably on a bag of oats, with the light turned off, and the room lit only by the flickering flames of the fire with the delicious smell of sausages cooking on the pan, I wondered if there could be a more comfortable place in all the world to talk or "tell" a picture.

What was home to me consisted of our street-fronted three-bedroomed house, an adjoining bakery and shop which we rented out to Mr Mullery, a farmyard where we kept livestock, and a house on the far side of the farmyard which we owned and rented out in rooms to usually three or four families.

My father worked in Dublin Corporation as a carter, driving his own horse and cart. He worked there for over thirty years on a contract basis, and in time, provided more horses and carts to the Corporation for which he engaged drivers. He also stored goods in his yard which we would collect and deliver. In short, his business activities in those days were the modern-day equivalent of the warehouse, storage and transport business. In addition, he kept pigs, sometimes born in our yard, which were fattened and sold to Donnelly's bacon factory, which was just 50 yards away.

My father was a perfectionist in his own way. Horses played a big part in his life and he would always look after them before attending to his own comforts, or those of anyone else, for that matter. The pig sties, horse stables, storage areas and yard were always kept clean and tidy.

My father probably knew everything that was to be known about working horses, and quite a fair bit about pigs. But outside of those animals, his knowledge of the animal world was very limited, and when it came to goats, he knew hardly anything at all.

My mother was a kind loving person who was generous to anyone with a sad story, and sometimes to those with no story at all.

My father was also kind. Never once that I can remember did he ask tenants in his rented house for their rent. They came in on Friday evenings with their money and the rent books. When the tenants, who were also our friends, fell on hard times, my Dad would not let them pay the rent. When one tenant with a wife and young child developed a serious illness, as a result of which he could never work again, my father allowed him and his family to live rent-free in the house for the rest of their lives. On the other hand, if I walked on his beloved corrugated iron shed roof, or did not attend to the many jobs around the yard assigned to me, or upset him in ways about which I knew he was "short-fused," well then I could expect a clip on the ear, and sometimes a lot more.

11

THE MAMMY AND THE DADDY

My father was the most honest, honourable, and generally decent man I have ever known. And I know that, in that respect, I can never fill his shoes. Indeed, just to be able to walk in his shadow is an everyday challenge for me. He believed in law and order, and justice in our courts. My father, even after his death, casts a long shadow.

If my father's greatest attribute was integrity, then my mother's was kindness. But let me tell you a bit more about my father. He was not just a Dubliner, he was a sixth generation Dubliner, born in The Liberties. He stood about 5'10", of square build. He had a big nose as well as a lovely big pair of floppy ears, which he dished out with gay abandon to nearly all of his offspring. I got the lot. His hair always seemed to be thinning, but he didn't look bald. He had two big sad eyes.

He was mostly even-tempered, but quick to anger if you transgressed in things that were important to him. He had a nice sense of humour in a very innocent way, in that he had half-a-dozen jokes, which down the years he told over and over again to any poor eejit who would listen to him, and he just loved to bang down the latch on our living-room door which made my poor mother jump to high heaven.

He was not an educated man, but was successful in a "street-wise" sort of way without being smart-alecky. He was liked, without being popular, but he was respected without compromise. If he had a failing, it was that he was too proud. He was affectionately known as "Master Pat".

My father and generations of his family were born and reared in the Cork Street, Ardee Street, Pimlico, and The Coombe areas of Dublin. His sister and brother still lived in Braithwaite Street when we lived in Cork Street. As my father could not get his tongue around the "th" of "Braithwaite," that street became known to him, and lots like him, as "Breffer

13

Street." Generations of my father's family had been jarveys, the equivalent of our modern-day taxi drivers. The horses and carriages were certainly more elegant than today's Mercedes 300 Diesel.

As I explained earlier, my Dad worked in the Dublin Corporation as a carter and his duties varied from bringing sand and cement, etc., to men working on road jobs, grassing and generally maintaining the roads around Rathgar and Rathmines. At one stage, my father was paid as little as eight pounds and fifteen shillings (that would be eight pounds and 75p) per week to cover his wages, and to feed and maintain the horse and cart. My Dad left home at 6 o'clock every morning, come rain or shine, for the thirty-odd years that he worked in the Corporation.

The pig business started slowly for my father and took a few years to build up. It was only when he managed to secure purchase of the offal from the Clarence Hotel, and I became available to work for him when I left school, that the business started to thrive. In the meantime, he bought other horses and rented them out to drivers who secured their own contract work.

He was a stickler for the care of his horses, their harness, drays and traps, and if those he trusted to drive his horses did not keep up his standards, then they would never again be allowed to drive them. He had a love for horses which I never encountered anywhere else, and to use a whip on one was absolutely out. We had only one whip in our stables, a long elegant driving whip with a solid silver ferrule on which my Dad's name was engraved. The whip was used only for show, when my father and mother went driving on a Sunday, in the two-seater back-to-back or round trap, pulled by his beloved and stylish black mare.

My father was not a religious man, but he was a good man. When tenants in his rented house came on hard times through illness and unemployment, he did not expect, nor would he accept rent. But he allowed them to maintain their dignity by giving them paint and wallpaper to decorate their rooms, telling them that their work in doing that would be as good as rent to him. In addition, my mother made sure that tenants who came on hard times didn't go hungry. But my parents were not alone in their concern for others. Other people on the street, even those in poor circumstances themselves, made sure that no one was forgotten or neglected. Gosh, how times have changed.

My father was not one to swear, tell rude jokes, or speak badly about anyone. But he was no saint either, for when he blew his top, and this he did from time to time about things which were important to him, he blew it well and truly.

I remember just two occasions when he swore, and swore so gloriously over the antics of a cat named Albert, that two women present blessed themselves, one fell over onto the ground, and some of the other onlookers gave him a standing ovation. But more about that later.

He once described himself to my mother as a "quiet" nationalist. But my mother, who believed that there would be no peace on this island until the last British soldier left it, had a less flattering name for his kind of nationalism. She was a Northern Ireland Catholic nationalist, and her strongly held views were anything but "quiet". Our home was not, however, a "political" home where great debates took place. My father felt that Mr de Valera would do whatever was necessary to straighten out the country, whereas my mother believed that Michael Collins was the greatest Irishman who ever lived, and with his death went all hopes of a united Ireland.

My father was "out" in 1916, and when recounting his adventures he made it clear at the outset that he was not in the G.P.O. He said that if everyone who claimed to have been inside the G.P.O. fighting for Ireland was in fact in there, the inside of the building must have been as big as the Phoenix Park. In 1916 my father was a jarvey, and his contribution to the Rising was that he took messages between the different Republican leaders. He maintained that his contribution to the Rising was more valuable, and his missions more dangerous, than firing a gun from behind sand bags. My father showed me a hard hat — standard head wear for the jarveys at the time — with a hole in it, which he claimed was a bullet hole, got as a result of a British soldier firing on him during one of his dangerous missions. The only mystery about the hole, as my mother pointed out, was that it was just a hole, one single hole. If it had entered the hat from the outside, it had not passed through the other side. When at times the Ma was having a flaming row with the Da over something, she would say, "I know now why that bullet didn't come out the other side, it got stuck in your thick skull."

My father's other great political interest was in the 1939-1945 war. He tuned in to every news broadcast on the radio, and in particular to William Joyce (Lord Haw-Haw) who broadcast from Germany. My father had his own inimitable way of pronouncing words, and, no matter how often my mother would correct him, he still pronounced them his way. On hearing of a Luftwaffe raid on London, he would tell us that "ten thousand bums fell on London last night". As a boy with a bright fertile imagination, I can tell you that the thought of being in London with all those "bums" raining down on me out of the sky, was a lot more terrifying than crashing planes or falling explosives.

When my Dad got worked up about law-breakers, or for that matter anyone who caused him a lot of upset, he would say, "If I had my way I'd put them up against the wall and shoot them." A few times he said that I too should suffer that fate, for some kind of desperate trouble I had caused. But more often when he lost his temper with me over something, he would simply call me a "yalla cur," because of my sallow complexion. But through everything, I knew the big slob loved me, and things he said in a fit of temper didn't mean that much.

My Dad imbued in me at an early age the fear of the Chinese. In many of the adventure films shown at the time, the Chinese were depicted as murderers and thieves, and it was many the nightmare I had about them taking over our street. When from time to time the Da would announce, "The yalla race will rule the world," it frightened the hell out of me. It was only when I "discovered" Errol Flynn that my fear of the Chinese went away, for after seeing Errol's exploits in *Robin Hood*, *Gentleman Jim*, and a whole lot of other action films, I couldn't imagine him being beaten by anyone. The Chinese may have lots of little "yalla" fellows with knives, but we had Errol Flynn. And if Errol ever got into a tight spot with the Chinese, there was always back-up from a lot of other favourite film stars.

My father was also a "soft" man, in that to my surprise I saw him cry a few times. He loved horses, and in particular the sport of horse trotting. His favourite film was *Home in Indiana*, in which Walter Brennan starred. It was the story of a trotting horse named Maudeen, and the tension was built up to the point, where, just before the biggest race of the horse's life, with the family fortune riding on it to win, it was discovered that the horse had gone blind. When Maudeen, despite the blindness, came through to win, the tears were streaming down the Da's cheeks. Three times he brought me to see the film, three times he cried like a baby. In the end he had me at it too.

He cried again, on another long to be remembered occasion. I was about seventeen at the time, and he was giving off to me about something I had done, which, he claimed, could have let down the family name.

I remember saying to him, "I am sick and tired of the bloody name of the Bolands, I wish to God I wasn't a Boland."

He just stood there in silence facing me, his sad eyes looking directly into mine, just as his tears started to come. He didn't say a word, he just turned and walked away from me. I was absolutely stunned and could do nothing but stand there for a few moments.

Then I went after him and said, "Dad, I'm very sorry. I'm very proud to bear your name, and I'll never dishonour it."

We hugged and "blubbered" a little. That moment will remain clearly with me for the remainder of my life.

My Dad's pride and joy was his black mare, which, he said, had been sired by a winner of the American Hambletonian Stakes, the most prestigious of all horse trotting races. No matter what her breeding lines were, she was a magnificent looking animal. My father used her almost exclusively for Sunday driving, in one or other of his traps.

During the week the mare was exercised either by the Da running up and down the street behind her with long reins on a martingale, or by myself, when I would yoke her up to the rubber-wheeled dray, and drive up to Jack Murphy's place in Kimmage to collect the Clarence Hotel offal. On some of these occasions, I would "let her go" for a mile or so, to brush out the cobwebs. As the mare came thundering down Clogher Road to the Sally's Bridge junction, people would stop cars and carts, get off bicycles, dash to front doors, stop walking on the pavement, just to see this magnificent trotting horse flash by.

If my father had known that I was driving her fast, he would have had my life. But the sheer thrill of being behind the reins of such a speed machine, was then, and is still to this day, one of the most exciting experiences I have ever had.

Once I let the mare fall when turning too fast and too sharply at the junction of Sundrive Road and Kimmage Road. I too took quite a painful fall, right from the top of the cart down onto the road. But my concern was for the mare, which just sat there on the road between the shafts, winded at least, but maybe more seriously injured. The road was slippery with frost which had contributed to the fall, and because of it, when she tried to get up, she was unable to get a grip on the road surface. But when I put my overcoat and jacket under her feet, and two onlookers did the same, believe it or not, she managed to get a grip and get up. The only injury she suffered was a pair of grazed front knees. Boy, was I relieved.

That was one situation where the Da might indeed have "put me up against the wall," had anything serious happened to the mare. Jack Murphy bailed me out by telling the Da he had witnessed the mare slipping on the frosty road. When I got home it was the mare that got all the attention, there being no interest whatsoever in my black eye, ripped knee, bruised shoulder, and a few bruised, possibly broken ribs. But I understood, and was happy that the mare was all right. After that I drove the mare most gently, for she was also my pride and joy, and I realised that she had been lucky to have escaped serious injury in the fall.

The Sunday drive was the big thing in my father's life. Preparation would start on Thursday night with the cleaning of the driving harness. The trap harness was black patent leather with silver fittings. After being cleaned and polished, it was carefully wrapped in soft cloths and left in the bedroom out of harm's way until "yoking up" time on Sunday. On the Saturday afternoon, the mare was washed down, had her mane plaited, and her shoes and pads checked. If any of the shoes or pads were loose, she would then be taken to the blacksmith for repairs. Before the mare could be returned to her stable, it had to be cleaned, washed out, and have new straw spread in it, so as to keep her clean for the next day.

On Sundays the mare was fed a little bit earlier than usual, after which she was checked over again to make sure that everything was in order. Then the cleaning commenced, using the dander brush, comb, and a lot of shining cloths. This brushing and shining of the horse would last for almost an hour, and then the hooves were oiled. After that the harness was fitted, and then she was yoked to the trap, which itself had been washed and cleaned up the previous day, and stored overnight in one of the sheds.

When the whole outfit was ready to travel, my mother would emerge with travelling rugs, sandwiches and flask, and get into the trap while it was still in the yard. Then the Da would lead the mare and trap out onto the street, where I would hold the mare by the winkers while he stepped into the trap through the door at the rear, closing it behind him. Then came the magic moment for which the Da had waited all week, when the mare, with ears pricked and high steps, moved off.

I have been to many horse shows over the years, including the R.D.S. shows, where I have seen winners in the Trap classes, but none of the winners could have competed with the excellence and style of my father's turnout. I can still see in my mind's eye my father and mother going off in the trap, the Da with his pipe clasped firmly in his teeth, his arms fully outstretched with the reins, holding back his champion trotting horse, as people all along the way stopped to admire the turnout. This was my father in all his glory.

The Fairyhouse Races on Easter Monday attracted a lot of horsy people from Dublin, their horse and trap turnouts being a colourful cavalcade along the main road of the Phoenix Park. For years on end, we joined the cavalcade on the Easter Monday, enjoying all the fun and excitement of the horsy people having their private races along the way, and shouting good-naturedly to each other as they passed by, with traps full of laughing women and children. My Dad was content to let his mare "prance" her way to Fairyhouse, showing off along the way, as other horsy men looked on enviously at mare and trap. But if he ever felt that he did not want a particular driver or horse to pass him, he would simply shake up the reins and our "racing machine" would move effortlessly away, with the Da grinning from ear to ear and onlookers cheering. When we got to Fairyhouse, he would stand by his horse all afternoon, while people gathered around to look at her and chat about how fast she could run.

My Dad's other love was playing cards, "twenty-five" being his favourite game. I too loved to play "twenty-five," but the game became dangerous for me when I discovered a way of dealing out the ace of hearts, whereby if I didn't have it in my hand, I knew the Da would have it in his. But the Da copped on.

In one particular game, when I opened my hand of cards to discover that I held the ace of hearts, the Da reached across the table, put his hand on mine, and quietly said, "If I discover that one of us is holding the ace of hearts again, which seems to happen nearly every time you deal, I will ring your yalla ears."

I folded my last two cards to give my father the game, in order to hide the ace. But when the Da suddenly reached over and turned up the cards to discover the ace before I had a chance to shuffle it into the pack, I immediately headed for the kitchen door, with a deck of cards flying after me, and my Dad's favourite war cry ringing in my ears: "You'll never be like the other fellow," he being my half-brother Wally, the perfect one.

Once an alarm clock followed me out the door with the same war cry, but he probably threw it to miss, just to show that deep down inside — sometimes too deep for comfort — he really loved me.

My father's one concession to drink was a bottle of stout which he drank every Christmas Day before his dinner.

My mother was less "colourful" than my father, but she was a beautiful, kind, and somewhat sad lady. She was younger than my father and slightly taller than him. But despite her good looks, I still ended up with the Boland nose and ears, while the search continues for the chin. She came from a farming family in Fermanagh, where she was always used to full and plenty. Whatever housekeeping allowance my Dad gave her, it was never enough, because of the kind of meals she prepared and her generosity to the neighbours who had fallen on hard times.

Because of both the Ma and Da's kindness to others, there was not too much spare cash around, and there certainly weren't any luxuries, whatever they might have been in those days. But we weren't quite as bad as the family in Maryland, where the daddy hung a rasher on the back of a door so that the family could rub their bread off it at dinner time, and that way not forget the taste of meat.

My mother was very good at looking after the animals around the yard, and although we were living in a Dublin farmyard, I feel that she would have been happier on a farm. She was a northern Catholic and a nationalist to her fingertips who had seen a member of her own family shot dead by the "B Specials".

When my Dad died and I took her to live with me in London for a while, she hated the place, and would say, "Don't ever bury me in this God-forsaken land whatever you do." In time we returned to Dublin where she died and is buried.

The Ma loved animals, particularly cats and dogs, and she could never bring herself to drown any of the many kittens and pups which were born in our yard. Instead she took the time to find good homes for all of them.

She married my father after his first wife died, leaving him with three daughters and one son, all of whom were fairly well grown up when she married Da. My mother was in her forties when I arrived, the only child from the second marriage, and "the shakings of the bag," as I was often described.

When I was about twelve or so, the last of my half-family got married, and I was effectively an only child from then on. In fact I was an only child from a much younger age, the others being so much older than me when I was born. Only one of my father's girls liked my mother and me, and I know that my mother thought the world of her. When she died suddenly in the prime of life, leaving a really nice husband and children behind, a great sadness took hold of my mother. I used to love to go to Ellen's house to play with her and her children, for it was a house of fun and love. When she died I lost a real big sister. On my father's death, virtually all contact with the rest of the half-family ceased. Years after both my father and mother died, I was persuaded to "reach out," as they say, to one member of my father's family, but my overture of friendship was not reciprocated.

I suppose that's the way it is sometimes with half-families.

TRYING TO FLY

The Wright Brothers and I had something in common, in that we both employed "aids" of one kind or another in our efforts to fly. Whereas the Wrights used an engine, wheels, wings, a fuselage, controls, etc., I used just brown paper, wood, glue, a rope, and internal combustion of the human kind. I did actually manage to get off the ground, but, unlike those of the Wright brothers, my flights were all downward from the top of a sand pit.

The adventure started on a summer's Sunday during school holidays. I was about nine or ten at the time, and was "taking the sun" on a big slab of concrete in our farmyard, it not having a blade of grass on which to relax. Seagulls often visited our yard to eat the pig swill, and as one hovered almost motionless overhead, I was struck by its effortless elegance of flight. I thought to myself "with a decent pair of wings I too could fly." And so I went into the workshop, my most beloved of all places in the yard, to make the wings.

The workshop was a very special place for me growing up. It had work benches, vices, and a wide range of tools, seemingly the kind to do any job imaginable. Also in the workshop was a big assortment of wood of all shapes and sizes, as well as pieces of metal, wire, nails, and screws. I could immerse myself from dawn to dusk for days on end in the workshop, making all kinds of weird and wonderful things. At dinner times my mother would bring out the meal to me, so I could eat as I worked, the alternative being I would not eat at all or do the "three-minute eating trick" at the table, the sight of which, my mother told me one day, she could no longer stomach.

I made the wings using thin laths of wood which I nailed, glued, and tied together at the joints with twine. Over this frame I glued thick brown paper of the type that bottles of stout were wrapped in, long before plastic bags were used. The wings were very carefully constructed from a design taken from my *First Holy Communion Prayer Book*, and modelled by a fellow named Angel Gabriel. Well, they were as carefully constructed as I could do it, but as the doctor was to remark later, one wing was definitely bigger than the other, and this might have affected the aerodynamics of the first flight and caused the resultant crash.

I had finished making the wings and was trying to figure out the best way they could be flight tested, when my pals arrived to see if I wanted to go to the sand pits, which were on the Tallaght Road, just above where the Walkinstown roundabout is today. I was considering jumping off the roof of our house to try out the wings, but as I was fearful that I might overfly the open space of the yard and land on, and possibly damage my Dad's beloved corrugated iron stable roofs, I abandoned the idea as being dangerous; dangerous, that is, from the point of view that my Dad might beat the fertiliser out of me if I landed on the roof. To test my wings at the sand pits seemed a much better idea.

In due course we arrived at the Tallaght Road. There were two ways to reach the sand pit. There was the safe way, by going around an inner field, but from the day that a "Beware of the Mad Bull" notice was posted at the gate of that inner field, there was, for us, only one route to take.

The stories of bravery and cowardice which followed the crossing of the "mad bull field" were many, and often told. To "funk" the bull and go around the safer way was a disgrace almost as painful as the injury we imagined the bull would inflict if he ever caught us. Young or old, big or small, fast or slow, were expected to "run the bull," and the fact that all of us escaped injury is just miraculous. I ventured the opinion once that the bull didn't really want to catch us, that he just enjoyed the chase. Another of my pals, whom we sometimes called Birdbrain, said it wasn't a real bull, only the farmer and his wife dressed up in a skin to frighten us away from the sand pits. I can tell you that Birdbrain sometimes frightened us more than the bull.

Having reached the inner field, and waited until the bull had turned away from us at the top of it, we then ran like the clappers, between the gate on our side and the one at the other side of the inner field, which as usual we reached just in time, the last of our legs disappearing over the top rail of the gate as the magnificent animal came crashing into it, determined to get us.

Having arrived at the sand pits, parts of which resembled a valley with sandy walls, some descending, some rising to in excess of 100 feet, I chose the highest point from which to flight-test the wings by gliding down into the valley. My friends and I, with but one exception, were all equally enthusiastic about the flight. We all agreed that the first test flight would be a simple glide, just to get the feel of the wings. The second flight was to be the one where I would do some tricks, like loop-the-loop, and dive-bomb from beyond and through the clouds. There was even a suggestion that one of the gang might come on my back for the third flight. But being a careful kind of kid, even at that age, and not wanting to do anything dangerous, I did not encourage the idea, being afraid that any extra weight could cause stress on the wing structure, thereby forcing us into a hard landing. Only a week or so earlier, I had seen a film in the Leinster cinema about test pilots, and that had taught me the importance of checking out everything very carefully before making a test flight.

The only one who was not enthusiastic about the flight was Birdbrain, who, in his own inimitable way of expressing himself said, "In the name of Jasus, don't tell me that you are going to jump off that bloody mountain with all that stuff tied on your arms. Sure you'll kill yourself."

Later on, as my pals gathered around my bed during visiting hours at the hospital, we thought of those words, and one remarked, "Maybe Birdbrain is not as big a gobshite as we think he is."

Half of my friends stayed down in the valley to witness the descent, while the other half came to the top of the cliff with me to witness the take-off. I tied on the wings, flapped them a few times to make sure they were secure, and did a few little vertical jumps, just to get the feel of movement off the ground. The fact that I didn't get "the feel" caused me no concern at the time, for I felt that once over the top, out in space gliding away to my heart's content, my only problem would be deciding when to come down. Little did I know at the time, that the decision as to when I should come down, or how I should come down, would not be mine.

At last the moment for take-off arrived. I went to the edge of the cliff, looked over the sand pit valleys and beyond to the mad bull in the field, deciding the direction of my first glide. Then I retreated about thirty yards, took a deep breath, and started my run.

In the film I had seen, the test pilots discovered any design faults towards the end of the flight. I discovered the problems in my test flight at the very beginning, in fact just as I left the top of the cliff, for I went straight down, just like a bag of coal.

As I landed in a heap of bone, flesh, and flight equipment, near to my friends below, who were terrified for a moment because they thought I was about to land on top of them, I realised that something had gone seriously wrong. It was not just the fact that the maiden flight had not gone according to plan, but it was also that I lay on the ground, under a heap of brown paper, wood, and string, unable to move. As my pals laughed themselves sick, those in the valley as well as those on top of the mountain, it was Birdbrain who ran as fast as he could towards the road to tell someone about the accident.

After what seemed like an eternity, an ambulance arrived, coming into the outer field as close as it could to the inner "mad bull" field, beyond where I lay in agony, surrounded by my laughing pals. As I lay helpless and furious on the ground, I threatened the hoors with what I would do to them if ever I walked again. This only made them laugh even more, and annoy me more.

Some of my pals had gone the "safe way" to the road to meet the ambulance, and told the two ambulance men with the stretcher, that the only way to reach me was through the mad bull's field. They didn't mentioned the mad bull, who was standing quietly behind the hedge. Halfway across the field, the ambulance men realised they had been fooled. They saw the bull at about the same time as it saw them, and a race started for the exit gate on the far side, both the ambulance men and the bull being anxious to get there first. Half my pals were on one side of the field, cheering on the ambulance men, and half on the other side, cheering on the bull. I could hear the cheering from where I lay, watched over only by Birdbrain.

My pals later told me, that apart from my take-off and landing, the race to the gate by the men and bull was the funniest thing they had ever seen. Unfortunately for me, the first thing the ambulance men did in their race to the gate was to drop the stretcher, which even in the circumstances was neither understandable nor forgivable, as a young seriously injured test-pilot was awaiting hospitalisation after his first emotional and bone-shattering flight.

One of the ambulance men reached the gate a second or so before the bull, but in his terror of trying to get clear of the top rung of the gate before the bull had his leg off, he overbalanced and fell heavily out over the gate, to safety if you like, but in doing so he broke his wrist.

The second man was not so lucky, the bull's horns spearing him in the arse just as he was hauling himself clear of the ground and about to jump the gate. According to eye witnesses, that jump was helped greatly by the bull tossing him about four or five feet over the top of the gate, before he landed smack on top of his colleague who was lying injured on the other side.

The cheers of my pals turned out to be too much for Birdbrain, for he took off, leaving me on my own, while he went to see what all the excitement was about. When he saw the carnage, heard the cursing and threats of the ambulance men above the laughter of my pals, and saw the angry bull with froth dripping from its mouth, he changed his mind about the farmer and his wife being dressed up as the bull.

The ambulance men had been badly injured, the fellow with the arse injury shouting for an injection which the other fellow could not give him because of his broken wrist. Eventually, limping and cursing, the two brave but battered ambulance men reached me. Both were as white as a sheet, the one with a broken wrist, and the other with a bleeding arse. Despite their pain, and the howls of laughter from my pals, they still managed to show sympathy for my predicament.

As the ambulance man with the arse injury — later to be called "The Arse"— examined me, his colleague — later to be called "Broken Wing" — was in great pain and unable to do anything but bend over me in agony holding his wrist, as my pals suddenly became silent, worried for me and my injuries. At the end of his examination The Arse announced that he wasn't sure as to the extent of my injuries, and he felt that I should be moved only on a stretcher. One of my pals suggested that Broken Wing, because he only had a wrist injury and could still run, should go back into the mad bull's field to get the stretcher. Both men lost their cool at this and made a swipe at him. Both missed, but Broken Wing overbalanced and fell on top of me to cries of pain from both of us. The Arse managed to straighten up Broken Wing with one hand, while at the same time holding his injured rear cheek with the other hand. Looking at the performance, I thought you wouldn't see better in a circus.

Broken Wing, though obviously in pain, nevertheless had a funny expression on his face, but I knew I dare not laugh. But when one of my pals said that someone should call an ambulance, I just burst out laughing, despite my own pain. At that stage, The Arse, having earlier learned as to how I came a cropper, told me that I was a stupid little bollix, and was more suited to go to a nut house than a hospital. Again the gang just burst out laughing, myself included.

Just then an elderly man came around the corner of one of the sand pits, wanting to know what was the matter. The Arse explained all, including their trip through the mad bull's field. The man then identified himself as the farmer who owned the fields. He asked the ambulance men if they were blind as well as stupid for not having read and taken heed of the notice about the mad bull in the field. He asked them if they had not seen the wheel marks through the grass around the bull's field where they could have driven the ambulance right up to where I lay.

At that point, my pals who had "guided" the ambulance men into the bull's field, visibly wilted under the stream of abuse from The Arse and Broken Wing, which was very descriptive and indeed magnificent in its own way, but still too strong for even our liberal young minds.

The farmer, ignoring the need for a stretcher, picked me up in his arms and carried me to the ambulance while The Arse and Broken Wing hobbled along together behind him, with the pals following, again laughing and teasing the two poor unfortunates, who forgetting their injuries, called the boys all the names imaginable, and indeed some not imaginable. The farmer tried to prepare the ambulance men for what was still to come by telling them that they should never have left any kind of vehicle in a field with cows. When we arrived at the ambulance, it was surrounded by cows and scraped and dented, quite badly in places. The sight of this on top of everything else made Broken Wing get sick. The farmer laid me down gently on one bed in the ambulance, and then helped The Arse to lay Broken Wing down on the other bed. The Arse absolutely refused to give the gang a lift towards home, but bravely undertook to drive the ambulance, or what was left of it.

When we arrived at the hospital and the rear doors of the ambulance were opened, there was a great deal of activity with a few nurses, a doctor, and one or two porters with a stretcher standing waiting for the "deliveries". Broken Wing — who should have been named "Broken Wind" because of his activities in the ambulance on the way to the hospital — was the one they seemed most concerned about, and he was moved very gently onto the stretcher. The farmer, who had travelled with us in the ambulance, remarked that he was the colour of death and wasn't long for this world.

The half dead Broken Wing, on hearing the remark as he was being carried out of the ambulance, came back to life responding, "Fuck you and your fucking bull and that stupid little bollix beside you."

The farmer's response was shouted after him: "Fanagans [the undertakers] will have you tomorrow. You'll be in the front page of the *Evening Mail* tomorrow night."

The next one to be removed, this time from the driver's seat, was The Arse, who was moaning and groaning as he was helped down onto another stretcher. Bad and all as he was, he still found the energy to shout "Little bollix" in through the open door of the ambulance as he was carried past the rear entrance.

The "little" part of the message suggested that it was for me, but the farmer obviously thought otherwise, for he shouted back his own message, "And you're a big stupid bollix for going into a field with a bull and parking your ambulance in a field with cows."

The farmer was very nice to me as I waited my turn to be removed from the ambulance. Ten minutes passed, and still no one came for me, as the farmer told me in terms then unfamiliar to my ears, what he thought about the ambulance driver and the hospital staff for leaving me there. Then he stretched out on the bed on the other side of the ambulance to await developments. Very shortly he was asleep, snoring to the high heavens.

Some time later, a big fat middle-aged nurse, probably a Sister, arrived at the door of the ambulance with more stretcher bearers and a nurse. When she saw the farmer laid out, she said, " Oh my God, there's two of them."

I was feeling a bit better by this time, the pain having almost left my legs and I was able to sit up on my elbows. I told the Sister that there was nothing wrong with him, that he had only gone asleep. She asked me if I knew his name, and I replied that he was a farmer.

Then she started gently tapping him on the cheeks saying, "Wake up, Mr Farmer."

Slowly the farmer began to wake up, and with eyes still closed, but with a big smile on his face, he reached out a big hand and touched the Sister, saying at the same time, "Oh Kathleen."

I thought to myself that if he opened his eyes and saw the face of your wan, he wouldn't be reaching out, and he certainly wouldn't be smiling at her. Kathleen, or whatever the Sister's name was, responded warmly, clasping his hand, and no doubt thinking that all her birthdays had come at once.

When the farmer woke up fully, he jumped back from Kathleen, as any normal full-blooded man would, first wanting to know why I was left waiting so long, and then giving his views in no uncertain manner as to the intelligence, vintage, and heritage of the ambulance men.

As I was being wheeled into the hospital, Kathleen walked behind me, telling me that I was responsible for two ambulance men being seriously injured, and a new ambulance being wrecked. She also called me a name and it wasn't "Pet," the name that most nurses seem to call patients.

The farmer, my great defender, on hearing the name she called me, told Kathleen what he thought of her and her staff for leaving a nice little boy like me lying in an ambulance in pain, while they all fussed over the "two stupid eejits," who had parked their ambulance in the middle of a cow field and then ran across a field with a mad bull in it, ignoring a sign that was as big as a house. He said that they were blind as well as stupid and they deserved all they got. He added that if his bull had been injured in any way, he would sue the hospital.

I was examined and had my legs bandaged in one room, and then I was removed to a ward. The sister in charge of the ward was young and gentle, smiled a lot, and had lovely blue eyes. When she spoke with a lovely country accent, she said, "So you're the test pilot. Tell me all about it, pet, how far did you fly?"

At that moment I fell in love. Looking back, I realise that my condition left me very vulnerable at the time and I really had no chance against those dreamy blue eyes, beautiful smile, and soft sexy voice. Life can sometimes be hard for a nine-year-old. Then and there, however, this older woman who smelled beautifully and 'crinkled' when she walked, was for me.

I started to tell her about the seagull and the Angel Gabriel design of wings in my First Holy Communion Book, but she asked me to stop while she got some other nurses and a doctor to come and listen. And as I went through the whole story of what had happened to me and the ambulance men, they just laughed and laughed. The farmer, who kindly stayed on to see if I was all right, also joined in the laughter. The doctor, picking up the remains of my wings, remarked that the Angel Gabriel must have had a disability, because one of my wings was shorter than the other.

Later that evening the Ma and Da arrived, having been told by my pals that I was in hospital. My mother was in tears, and as usual just wanted to hug me. I must have been very huggable then. The Da just wanted to know what had happened. Blue Eyes, perhaps noticing the look of fear which the Da's question had brought to my face, told him that I had "fallen" when playing in the sand pits. Gosh how I loved that woman. She told him that my legs were sprained but that I would be out, fit and new again, in a couple of days. The farmer, catching her eye, told the Da that I had indeed fallen and that he had helped me. Shortly after that, the Ma, the Da, and the farmer went home, and with Blue Eyes crinkling and smiling over me in the bed, I went to sleep.

I awoke the next morning excited about the prospect of my second flight, for during the night, in the many dreams or nightmares I had about my first flight, I discovered, or thought I had, what had gone wrong the first time. But before I could even throw a leg out of the bed, my mother was at my side, hugging me and kissing me, and crying all over me, calling me her little son. After I assured her I was all right, she left to do the shopping.

Blue Eyes came to say that she was going home, but would see me later that evening when she came back on duty. My departure was further delayed by the arrival of breakfast, which I gobbled down quickly. After that I asked for my clothes, which they refused to give me at first.

Then Kathleen arrived, and after listening to the arguments about the clothes, she settled the matter with one short sentence, "Give the little trouble-maker his clothes and let him go."

Within the hour I was back in my beloved workshop, where I started in to making a new and better pair of wings. When my mother arrived home from her shopping, to hear the sound of hammering coming from the workshop, I told her that the Sister had given me my clothes and said I could go.

One of the problems I had encountered when I jumped off the cliff was that the air forced my two wings up into a vertical position, and therefore, being unable to flap them, I immediately crash-dived instead of just gliding gracefully away.

At least that's how it seemed to me after I had slept on the problem. I figured that the way to correct the design fault was to make sure that my wings were not driven up vertically again by the upsurge of air. This design correction was to be accomplished by tying two ropes to both wrists, and then tying them to my ankles. Then, when the air pressure came on my wings, my wrists, held down securely by the ankle ropes, would keep the wings horizontal, thus allowing me to glide, and later do a few loop-the-loops.

I had just completed the new set of wings when some of my pals arrived to see if the Ma wanted anything brought down to me in the hospital. No doubt they were hoping for cake, lemonade and bags of toffee which, knowing the gang, would never have reached me in the hospital. They were surprised to see me, but even more surprised when I told them that I was leaving for the sand pits again, to try out another pair of wings. Without Birdbrain present to pour cold water on the project, my pals' optimism and confidence for the success of the project, soon rivalled my own, and so we all set forth for the second test-flight.

On the way to the bus stop however, Birdbrain appeared, and with a shout of, "You're all bloody mad," he nevertheless joined the throng.

We arrived safely at the sand pits, quickly fitted the wings and tied on the wrist and ankle ropes. Now I was ready for my second test flight. With half my pals below in the valley, and the other half on top of the mountain with me, I started my run.

It's not easy to run fast with ropes tied to your ankles and wrists, nor, I am told, is it easy to look at such a spectacle. I did not so much glide off the cliff top, as tumble off it. And whereas in my first flight I managed to propel myself some fifteen to twenty feet out into space through being able to run fast without the hindrance of the ropes, on this flight I managed only to crash into the wall of the sand pit itself, bouncing three or four times off it as I fell. This second landing was far more painful than the first one, but my pals still managed to get a good laugh out of the performance.

This time my injuries were more serious than the first time, for my legs, arms, and back were hurting me terribly. Totally winded, I closed my eyes and lay still.

"He's dead," said one of the lads.

"Someone better get an ambulance," said another.

Then Birdbrain, always one with the great one-liners, said, "It's not a bleedin' ambulance he wants, it's a hearse."

I opened my eyes, and for just a second saw the concerned look on my pals' faces before they started to laugh again, one of them saying, "Look, he's alright, he was just actin' the maggot."

Birdbrain duly did his trick and got an ambulance, this time bringing it in the proper way. The ambulance drove straight to where I lay, and I was lifted into it by stretcher. When the boys asked if the ambulance men would give them a lift towards home, they were told to get in.

These ambulance men were far more efficient than the previous day's crew, but then of course they did not have to contend with a mad bull, broken wrist, speared arse, and a wrecked ambulance.

The ambulance moved smoothly along with one of the ambulance men kneeling by my bedside, with my pals sitting in a row on the other bed. I was just lying there in pain unable to speak, just crying.

The ambulance man was concerned about me and held my hand, saying from time to time, "Don't worry son, you'll be all right." Then he asked, "What happened to him?"

But before I or anyone could answer, he went on, "We had another ambulance out this way yesterday. Some eejit of a kid tried to fly and nearly killed himself jumping off a cliff. And there was a crowd of hooligans with him, who ambushed and nearly wrecked the ambulance, and then they turned a bull loose on the ambulance men. One of them had his wrist broken by the bull, and the other one was gored in the …" Here he hesitated before adding "in the thigh."

At that point a voice spoke up from behind him, "It was in the arse, mister, the bull stuck his horns in the man's arse. I saw all the blood pouring out, and he was holding his arse all the time. It was deffney his arse, mister."

And as the ambulance man looked around in surprise, another voice perked up, and to my horror said, "And that eejit of a kid you were talking about mister, that's him lying in front of you. He was flying again today."

And before the poor ambulance man could sort out his mind, another voice said, "And mister, we're the crowd of hooligans that was there yesterday. We were there again today."

By this time I was lying there helpless and terrified, hardly able to believe what I was hearing, and sure that the ambulance man would at least strangle me, apart from whatever he would do to them. All I could do was cry more. Then the ambulance man, still reeling from the information just given to him, heard the true story of the previous day's happening, given in dribs and drabs from the lips of the hooligans.

"Mister, they left the ambulance in a field with a whole lot of cows, and it was the cows that wrecked the ambulance."

"Mister, there was a notice up to stay out of the field because of the mad bull, but they went into the field for a short-cut."

"Mister, mister, the one with the broken wrist, Broken Wing, fell off the gate, the bull never touched him."

"And mister, the bull got the other one in the arse just as he was nearly over the gate. It was great fun. It wasn't our fault, mister, the ambulance men were just stupid."

The ambulance man might not have got a true story of what happened, but he got it as close to the truth as the boys were prepared to give it.

After all this information had poured out, there was silence, except for my sobbing. The ambulance man was speechless. He just looked at the now worried faces behind him, and back down at me sobbing. After what seemed like an eternity, he quietly said, "Broken Wing, The Arse." Then he smiled and added, "Do you kids do this sort of thing often?"

Then Birdbrain piped up. "The last time he did something stupid was when he made a submarine and nearly drowned."

And then someone else added, "And one time he tried to hypnotise an Alsatian watch dog and got bit on the hand."

"Hold it, hold it, hold it," shouted the ambulance man. And we all silently waited for his reaction, as he scratched his head, rubbed his eyes, gave a big sigh, and said, "Holy Mother of God, tell me I'm dreaming all of this." Then he looked down at me, took my hand gently and said, "Don't worry son, you'll be all right."

After a time of further silence, Birdbrain, not content to hear the man saying I would be all right, asked if I was going to die.

Then someone else said that they hoped I would live until they went back to school after the holidays, so that they could get the day off for the funeral.

"Well," said the ambulance man, "it's an ill wind that doesn't do somebody good. The two matadors will be out of work for a few weeks, and that means that Charley and me will be able to do lots of overtime."

The boys were dropped off near Dolphin's Barn, promising to tell the Ma where I was going. When the ambulance arrived at the hospital and the doors were opened, there stood the bould Kathleen. At first she did not recognise me, covered right up to the neck with a blanket.

"What have we got here?" she asked the ambulance man, as he and the driver began lifting me out of the ambulance on the stretcher.

"This little fellow fell into a sand pit when he was playing with his pals," lied the ambulance man.

Then Kathleen recognised me.

"It's that little trouble-maker we had in yesterday. Is the ambulance all right?" she shouted, taking off around the front to examine it. Having satisfied herself that the ambulance was all right, she then walked by my side, still giving out about me as I was carried into the hospital once again.

The doctor who was on duty the previous evening was there again to examine me, I remember that this examination took longer than the previous night's. In the middle of it, the Ma and Da arrived outside of the cubicle. I could hear my mother crying, and the Da asking about me.

After a time they were allowed in to see me for a few minutes, before I was brought into the proper ward. There my own Blue Eyes greeted me. I still couldn't, or wouldn't speak, and the tears still came. Blue Eyes gently dried them, gave me a little hug, and told me not to worry as there was nothing seriously wrong with me, and I could go home in a few days.

Then Kathleen appeared beside my bed and stood silently looking down at me. After a time, she put her hand gently on mine and said, "Nurse, look after him. He's a gutsy little fellow, and he's the first flyer we've had in the hospital."

My only other attempt at "space travel" was when I jumped off the roof of our house with an open umbrella, intending to glide, parachute-fashion, down to the ground. Unfortunately, this was another experiment that went wrong, although the only pain suffered was when the Da gave me a few belts for turning his best umbrella inside-out. After that, I sort of came to believe that if God had intended me to fly, he would have let me grow a proper pair of wings.

THE CAKE SHOP

For as long as I can remember, Mr Mullery rented the building next to our house from us, which he used as a bakery and shop. He had returned from America with his wife, two daughters called Annie and Sally, and a son called Harry. Mr Mullery's greatest worry in life was his wife's health. She was a frail little woman, but he was not exactly the picture of health. In the end, as is often the case, she outlived him.

I don't know why it was called "The Cake Shop," as all he made was griddle loaves, plain or with currants, and pancakes once a year on Shrove Tuesday.

Mr Mullery was a gentle, quiet-spoken person with a nervous disposition. He worked very long hours, usually starting at 5 am, and working through until 5 pm. The bread was baked on an enormous open-plate oven which was heated by coke. This fuel gives long concentrated low-flame heat, and during the war years it was the only fuel Mr Mullery could get , and then only under licence. When the day's baking was finished, he would "settle-down" the fire for the night. The first job each morning was to get it glowing hot again. On Saturdays, he would riddle out all the ash and cinders, and reset the fire, ready for the coming week. At the same time he would wash down the oven, and generally tidy up the place.

The only part of the bakery which never seemed to get cleaned was the floor. It had a permanent covering of flour, dotted with currants and other baker's debris which, over the years, built up to a depth of at least an inch.

Mr Mullery was not only the baker, he was also the manager and shop assistant. He would sometimes drop money on the floor, losing it in the floury mess underfoot. Years later, when Mr Mullery died and the bakery closed down, my Dad set me the job of cleaning the floor. I wheeled barrow loads of the compressed floury substance out to the yard for dumping.

After finding first a sixpence and then a shilling, I set out to riddle every piece of the material, and accumulated a fortune. Just as I had made out a list of all the things I was going to buy, Dad told me to take all the money up to the Mullery family, that it was theirs. My appeal to the principle of "finders keepers, losers weepers" fell on deaf ears.

When Mr Mullery arrived for work, he would change into white trousers, a white short-sleeved shirt, white hat and white slip-on shoes. This outfit, together with his pale skin, made even paler by the constant haze of flour, gave him a ghostly appearance. On more than one dark morning, Mr Mullery unwittingly terrified the passers-by by stepping innocently onto the street for a breather.

Mr Mullery mixed his ingredients in large bowls. Then he transferred the mixture onto the worktop, beside a heap of flour. Next he would fill the top of the oven with metal baking rings. Then, moving fast, he would cut out rounds of dough, roll them in the dry flour, and press them into the baking rings, spreading them out with his bent knuckled fingers. When all the rings were filled, it was nearly time to start back on the ones he started with, quickly running a knife around the inside of each ring to free the bread. Then, because the rings were hot, he would flick them up from around the bread, and off the oven top and onto the floor. Next he would turn the bread over to cook on the other side. When both sides were done, he would place two metal blocks on the oven top, and stack the rounded breads on their edges between them, turning them from time to time to cook and seal the edges.

When the first batch of loaves came off the oven, Mr Mullery set about gathering up the baking rings which he had earlier tossed on the floor. Bits of dough were scraped off them, and the whole process would start all over again.

The loaves came in just one size, about 10" round and 2" deep. Mr Mullery let us have as many loaves as we could eat, for free. When piping hot, they were delicious split across the middle and smothered in butter and jam, which melted into the bread. But like a lot of things which come free, you soon get tired of them, and for a time, a black crusted Boland's loaf would knock spots off Mullery's bread for taste.

Mr Mullery's son Harry worked with him, delivering to

shops and restaurants on a messenger bike with a basket on the front. As the business expanded, Mr Mullery bought a second bike, and took on a young fellow named Matt Canavan to help with the baking and the deliveries. When the business expanded further, Matt worked full-time with Mr Mullery in the bakery, and Harry graduated to a little delivery van. Luckily, however, they kept the bikes which came into their own when, for one reason or another, the van was off the road.

There was one good reason, involving myself, when the van had to go for repairs.

One day, when I was about ten, I drifted over to where Harry had parked the van, at the top of the sloping yard. In the spirit of scientific enquiry, inspired by stories of Galileo and his experiments, I decided to see what would happen if I reached in and released the handbrake. The results were spectacular. The van rolled forward, slowly at first, then gathering speed at a great rate. I stood open-mouthed as it smashed through Dad's beloved gate and hurtled across the street.

The van was in no condition to do any work for a few days, nor would I have been if the Da had got his hands on me. I made myself scarce for the next few days while the Da made good the wreckage I had caused. By the end of the week, a new gate was in place, the van was repaired, and I was forgiven. After that, however, whenever I tried to get into the van to mess with the gears, I found it was locked. It was as if no one trusted me …

Pancake Tuesday was a special time for Mr Mullery, for he made hundreds of pancakes for his customers. He would work on them all through the Monday night. I used to beg the Ma and Da to let me stay up to help him. I liked the work and I could eat my fill of pancakes, but, best of all, I got to stay home from school the next day. Mr Mullery welcomed the bit of help, so I usually managed to swing it. After the first few hours, during which I stuffed myself with pancakes, washed down with a few pints of buttermilk, the appetite was satisfied, and after that it was just work and staying awake.

We always had gallons of buttermilk, because even if the Ma wasn't making butter, we stored churns in our yard for Mr Waldron, a carrier down the street who let us have as much of it as we liked. Buttermilk was a popular drink, and if you had it with hot new potatoes covered in butter, it made a first-class meal. Waldron's would on occasion ask us to get rid of it for them, and then we would invite the neighbours to bring in their milk cans and help themselves. Sometimes, when everyone had taken all they wanted, we fed the remainder to the pigs.

Mr Mullery's elder daughter Annie made delicious toffee, which she got her Dad to sell in the shop. The toffee was sold in bags for a penny and two pennies. Annie did a roaring trade with the boys and girls passing up and down to the many schools in the area. Even though I let my pals think I had unlimited amounts of toffee, I seldom got any for free, Annie's stock-control system being very tight. This didn't worry me too much, for even though it was very tasty, a little went a long way. On Pancake Tuesday particularly, I was the envy of my pals for having access to both the pancakes and the toffee. And when you added in Joe Mullarkey's icecream, which he made in our yard, the envy knew no bounds.

Mr Mullery was an educated man by our standards, and the Ma would often say, "Let's ask Mr Mullery, he's sure to know." One thing he certainly did know about was long-division sums, for whenever I needed help with this kind of sum, which bugged me even more than the others, he would always do it like lightning.

When I was growing up, there was a formality with people's names which is now a thing of the past. Whereas today there is a "get on Christian name terms as soon as you can" attitude, in those days, people were nearly always addressed as "Mr," "Mrs" or "Miss." We applied these titles as a matter of course to most of our neighbours, and to the various people whose business led them to our yard. Only very close friends and relatives were addressed by their first names. And although my father was known in some of the horsy circles as "Master Pat," nobody called either him or my mother directly by their first names. On the other hand, we little people were called all kinds of names, most of which we deserved.

My sister Ellen did not often get into trouble, but I once witnessed the white heat of my Dad's anger directed at her. She and her pal dressed up as ghosts, and, sneaking down to the bakery at 6 o'clock on a dark, quiet winter's morning, suddenly jumped out screaming at poor Mr Mullery, who promptly fainted. Even when he came around, he was too weak to continue working that day. But he recovered and the girls apologised, chastened by a furious tongue-lashing from the Da.

During the war years, although fuel was in short supply, Mr Mullery still got his coke. It was a valuable commodity, which my Dad allowed him to store in a shed. This shed, or rather the coke in it, was once the target of a midnight raid. The thief was one of our tenants, whom the Da regretted letting in.

Rover had not yet entered our lives, and my Dad maintained that the watchdog we had was not much good. The raider managed to get into the shed and collect a bag of coke without alerting the dog. He was not so lucky on the way out, for the dog heard him climbing over his back gate and, suddenly realising his purpose in life, launched an attack. The raider dropped the bag of coke and fled, but not before the dog had ripped a piece out of his trousers and attempted to do the same to his bum.

The next day, having had the wound treated in hospital, the tenant complained to the Da that when he had gone over the gate to get his son's ball, the vicious dog had attacked him, and he wanted compensation. My Dad's only comment, looking down at the drowsy old dog, was, "If I could believe that you bit this man, and at the same time carried over a bag of coke and dropped it outside the gate of his house, I'd have a medal made for you."

When Harry Mullery arrived at the bakery in the morning, there would be a great flurry of activity, as the first of the orders were got ready. He leaned the messenger bike against the outside wall of the bakery, and carefully packed the orders in the large basket. When the bike was loaded, two people had to hold it steady as Harry climbed into the saddle, and started to pedal. After a few initial wobbles, Harry would get the bike straightened out and go merrily on his way.

As he dropped off orders along the way, the load would lighten, making it easier for him to restart the bike each time. When the basket was empty, Harry would race back to the bakery for the next batch of orders. Sometimes he would arrive back waving his order book shouting, "Extras, extras," which was a signal for Mr Mullery to start baking extra bread. Some days the extra orders would be light, and Mr Mullery would get home early to be with his sick wife, and catch up on some sleep. When this happened, Harry or Sally would mind the shop. They did quite a good business in bread and toffee.

Despite his best efforts, Harry often crashed the heavily laden bike. When Harry was seen bandaged, or hobbling around the place, Mr Mullery would explain, "Poor Harry crashed again."

The reason was that he usually had difficulty in getting the bike moving after making a delivery or stopping in traffic. Although the basket had a leather cover to protect the bread, Harry often crashed so badly that the cover flew open, scattering loaves on the road. When this happened he would have to pick himself and the bread up and go back to the shop for a fresh load.

On the days that Mr Mullery went home early, leaving Harry in charge, the gambling would start. Harry would open the racing page of the newspaper, and after studying form, would write out his doubles, trebles, and accumulators, and send me up to Kilmartin's to place the bet. As I was too young to go into the bookies myself, I would ask some of the men standing around outside to place it for me.

By the time I got back to the shop, Harry would have whipped up a big jug of pancake mix, and with pancakes, mugs of tea, and as much toffee as we cared to eat — his sister Annie's strict bag control systems fell apart when Harry went to work on them — a great party ensued. If Harry's horses won he would be in great humour, laughing and joking. He would stand at the door of the shop, and shout to passing acquaintances, "I got up a nice double today."

My share of his winnings would be the price of the pictures, or a bag of fish and chips.

But if Harry lost, things were different. You wouldn't hear a word out of him. He would become very quiet, almost sullen, concentrating even harder on his work, and sometimes snapping the head off customers. The following week, when Harry got another week's wages, there would be more gambling on the horses. But he was not a real gambler, for his weekly investment was modest. If he won he would celebrate, if he lost he would mourn for a day or two.

Mr Mullery had good and bad customers, and whereas he

would always negotiate with even the worst of them, continuing to supply bread while awaiting payment , not so Harry. Harry had less patience, and I would say he was more realistic than his Dad.

On one occasion when I was alone with Harry in the shop, a bad debt customer arrived in with an armful of loaves, shouting, "These are stale, how can I sell stale bread?"

"They weren't stale when I delivered them to you last week," retorted Harry.

"Last week nothing," came the reply, "you delivered them only this morning."

Harry stuck to his guns and the shouting match continued until finally Harry lost the head, picked up one of the loaves and smashed it over the customer's head, bellowing, "Yes, you're right, it is stale." Before the man had a chance to react, Harry had whacked him with a second loaf, roaring, "This one is too." The customer fled, yelling abuse. Harry leaped over the counter, and armed with a pile of loaves, took off down the street after him, breaking chunks off and flinging them as he ran, shouting, "Here's more stale bread."

Naturally a crowd gathered, and we all gave Harry a big cheer when he returned triumphant from the chase. The hero smiled hugely. He acknowledged the cheer, with his hands clasped over his head and did a little dance around the pavement, just like a boxer in the ring. Harry enjoyed the incident, and made more tea and pancakes to celebrate and broke open more of Annie's toffee bags.

Harry took a great interest in the nurses from Cork Street Hospital, who often queued at the bus stop opposite the bakery. Sometimes he would wrap up a loaf of bread and a couple of bags of toffee in brown paper, and say to me, "Here, give this to the one in blue and tell her that Harry said he loves her."

These gifts were always graciously received, sometimes with a return message:

"Tell Harry I love him too."

I don't know how often, if ever, Harry followed up on these parcels. Some of the nurses used to visit the shop, but he never told me any details.

When I was quite small, I couldn't see much difference

between the objects of Harry's affections and their less beautiful colleagues, and once I made the mistake of giving the precious parcel to the wrong person in the queue, an older lady with a moustache. On getting the message of "love" she hastened eagerly across the road to meet Harry. Harry, seeing the mistake I had made, disappeared under the counter. He put his finger to his lips, cautioning me not to give away his hiding place, and shook his fist at me. When the lady left, promising to return, he got so annoyed at me that he told me I was blind as well as thick. He threatened not to let me deliver any more parcels, or put on his bets. I promised to be more careful in future, not wanting to miss out on the fun.

Mr Mullery eventually died, and though Harry and Matt kept things going for a time, Harry's heart wasn't really in the business. He went to Wales to work in the mines, and came back only once, for a holiday, all dressed up and waving wads of cash. I never saw Harry again, and more's the pity, for he was one of those people who adds sparkle to life.

THE PICTURES

Going to the pictures (the cinema), was a big thing for us, both as children and as adults. It was pure escapism from everyday worries, of which, believe me, there were plenty. The cinema was the gateway to a magical world, enjoyed by young and old alike.

In the same way in which, for example, sport — with its present-day massive television exposure — is discussed, so also were the pictures when I was growing up. It was not unusual to go into a shop and find a group of people excitedly discussing or "telling" a picture. Our local cinemas were the Leinster and the Rialto, but occasionally we ventured across to the Lyric in James's Street, or down to the Tivoli in Francis Street. Very occasionally, usually only when we were desperate to see a particular film which we had missed at our locals, we would venture into the "Mero" in Mary Street, and one or two other cinemas on the north side.

It was fairly safe going to the Leinster or Rialto, but going to the Tivoli could be risky, for there were some tough gangs around Carman's Hall, The Coombe, Thomas Street, and Francis Street. Besides the pictures, our entertainment as kids was mostly on the streets. Kids went around in gangs which encountered each other continually, so the opportunities for getting into fights were all too plentiful. All that was needed to start a fight was for a member of one gang to say something in passing to a kid in the other gang, and pride demanded that the second kid retaliate with more words. Before you knew where you were, fights were being arranged.

"You stand out to him, I'll stand out to you, that fella over there can stand out to this fella over here," and so on.

Or, with one mighty flurry of arms, legs, elbows, and a whole lot of shouting, both gangs got stuck straight into each other. A feature of the fights was that there was an unwritten law that only fists were used. To kick, or to butt with the head, was unheard of, and indeed on one occasion I saw a gang of real tough kids turn on one of their own in a fight because he tried to kick his opponent. This "Marquess of Queensberry" attitude in fights continued right up into my teenage years and beyond.

The pictures were a great influence on us as kids, particularly in the matter of fighting, when the good guys and the crooks always fought fair, even when they fought desperately for their lives on top of trains, planes, and high buildings. The fights we got into as kids usually lasted only a few minutes, and finished when one of the fighters was getting too many clatters, or sustained a bloody nose, or when a passing adult would run at us, sometimes shouting and pulling us off each other by the ears. Believe me, there were times when I was glad to have an adult pull me and my opponent off each other by the ears. My ears seemed particularly suited for getting a good grip.

My favourite male film stars were Errol Flynn, William Boyd (Hopalong Cassidy), Tyrone Power, Sidney Toler (Charlie Chan) and Buster Crabbe. When any of these appeared in a picture, you could be sure of action. I also had favourite female stars, and boy did they cause excitement. One of them was the Hollywood sex goddess Rita Hayworth. When I saw her dancing in *Gilda*, she aroused in me hitherto unknown but wonderful feelings, feelings which my conscience eventually forced me to share with a priest, for they seemed always interested in that sort of thing.

If I had a penny for every time a priest asked me, "Did you touch her? Did she touch you? Did you take pleasure out of it?" I would have had quite a few bob.

When I eventually confided in a priest about the "wonderful" feelings that Rita Hayworth aroused in me when she danced, I heard a sharp intake of breath — which for a moment made me think that he too had seen her dance in *Gilda* — then he asked, "Did you entertain bad thoughts?" You bet your sweet life I entertained them.

Another film star I liked was Veronica Lake. I suppose I felt sorry for her because she had only one eye — or so I thought at the time. But when I discovered that there was a second eye underneath the beautiful blonde hair which hung down over one side of her face, I still liked her. Jeanette McDonald was another actress I fell for, mostly because she reminded me of Julie, my first real love, but partly because, after seeing her in *Rose Marie*, I fancied myself sitting on horseback on top of a mountain in Canada, wearing a red coat, a funny hat, carrying a gun, and singing love songs to her across a valley as she sang them back to me from another mountain top.

Later on I fell for Greer Garson, and later still for Doris Day. When Marilyn Monroe appeared on the scene, I was big and the clock had turned full circle and it was back to the Rita Hayworth days, the wonderful feelings, curious priests and their questions as to whether or not I had entertained bad thoughts. At this stage in my life, when Marilyn was beginning to display her particular talents, the bad thoughts were entertaining me.

I fell hopelessly in love with Rita Hayworth when I was about twelve or so, and I wrote to her in Hollywood telling her so. While I was waiting for a reply, I discovered Veronica Lake, and she won my heart. I then wrote and told Veronica that I loved her. I confided this to Eileen, a girl of my own age who lived down the street . She said that I should straight away write to Rita and tell her I had fallen for Veronica, otherwise the two of them would be over to get me. I wrote to Rita and gave her the bad news. It was a bit hard, for every time I thought about her dancing in *Gilda*, I fell for her again. I posted the letter to Rita on my way down to Clarke's butcher shop in Meath Street to get some pork chops for the Ma.

Years later, I found out that Prince Aly Khan, a direct descendant of the Ismaeli prophet, and heir to the fabulous wealth of the Aga Khan dynasty, also saw Rita dance in *Gilda* — a private showing of the film, no less!

He possibly got the same "wonderful feelings" about her that I got, but he was on the spot in Hollywood, and in a position to pursue her more easily than I. Eventually they got married. I suppose it's as well that I wrote to Rita withdrawing my offer of love, for, faced with the choice between Aly's incredible wealth and experience, or my youth and magnetic animal sex appeal, the poor girl would have been in a terrible state, not knowing which of us to choose. Sadly, though, the marriage did not last long, and I often wondered was it because I had let her down, leaving her too heartbroken to share her love and her life with anyone but me. I don't suppose I will ever know now.

When passing by the end of St Margaret's Terrace in Pimlico, on the way to Meath Street, I bumped into Julie. She was a beautiful young girl of about my own age, and all of our gang, and all of the Connolly gang from "The Tenters," were in love with her. She had never previously given me her time, when my gang and I followed her gang around. Julie and some of her friends from the area went to the Holy Faith Convent in The Coombe.

Julie spoke to me, but I could not concentrate all that well on what she said, for I was just soaking up the beautiful music of her voice, the little laughs which interspersed her conversation, as well as drinking in her smiling lips, shining eyes, and flashing white teeth. All this drinking in and soaking up made me fall hopelessly in love with her, and this love affair, this one-sided love affair, was to last for many a long year. When Julie let me go on my way, after first giving me a bite of her Mars Bar, I immediately turned around, forgetting all about the pork chops, and went home to write to Veronica breaking it all off and telling her that Julie was going to be my love for always.

On the way back down to post the letter, I bumped into Eileen again and told her all about Julie and my new letter to Veronica. With that she burst into tears and told me that she loved me. For a twelve-year-old or thereabouts, this was too much to handle. Being up to your armpits in girls is one thing, but when they start crying at you, that's a white horse of a different colour. Eileen didn't mind me talking about or writing to girls in Hollywood, because she probably felt they wouldn't be found dead in Cork Street. But when I mentioned the gorgeous Julie, that was too much for her. Her tears drove me at full speed to the post office to get Veronica off my back, and at the same time I resolved never again to play Doctors and Nurses with Eileen.

When going through Meath Street or Thomas Street, it was my practice to go into the Chapels and say three Hail Marys for the poor souls in Purgatory. When on this occasion I went into

Meath Street Chapel, confessions were being heard, and in a moment of weakness, I felt that I had to tell about the way I had treated poor Rita and Veronica.

I had a moment of foreboding just before I entered the confession box, when I heard the priest shout out in a voice which could be heard all over the chapel, "You did what? Holy Mother of God!"

Then the boy in confession burst into very loud crying, and the priest shouted even louder, "Get out, get out, and tell your mother to come around to see me straight away." So much for confession being about talking directly to God …

I was petrified with fear when I entered the box, waiting for the priest to slide over the little door when I could tell my sins. When he did slide it over, I began with, "Bless me Father for I have sinned. It's a month since …"

At this point I was interrupted by a shout from the priest which everyone outside the confession box must have heard.

"A month? Did you say a month? What kept you away for a month?" he asked.

I told him that I had no sins to tell him until today.

"Go on," he said.

I told him that I had told Rita and Veronica that I loved them. Then I told them that I loved Julie instead. I told the priest that Eileen started to cry when she heard this, and told me that she loved me.

After telling the Priest all this, there was a terrifying silence for a time, before he asked, "Are you a little pervert?"

The nice friendly way he asked the question put me at ease, and then I calmly answered, "No Father, I'm a Boland."

All hell seemed to break out with my answer and he flew into a rage shouting, and using words I could not understand. Nor apparently could the people outside understand them, judging by the looks on their faces when I eventually fled the confession box.

But, meanwhile, the priest, obviously anxious to know more about my relationship with the girls, pushed on.

"What are the names of these girls?" he asked.

"Rita, Veronica, Julie, and Eileen," said I.

"Did you touch them?" asked he.

"I didn't touch Rita or Veronica, but I did touch Julie and

Eileen," I confessed.

"Where did you touch them?" asked he.

"In Pimlico and Cork Street," said I.

"That's not what I mean," shouted he at the top of his voice. "Did you touch their bodies, did they touch your body?"

I was becoming terrified at this stage, and the thought of having to go out of the confession box and meet all the people face-to-face in the chapel, and maybe others from half a mile away who couldn't have helped overhearing the priest, made my stomach churn and my legs shake.

In a now high-pitched shaking voice, I said, "Julie put her finger in my mouth when she gave me a bite of her Mars Bar, and when I was playing Doctors and Nurses with Eileen, she …"

At this point, just as I was going to tell him the best sins he'd heard all day, he suddenly jumped up out of his seat and shouted out very loudly, "You are a little pervert."

All this shouting, and now the jumping up bit, was too much for my fast-fraying nerves, and as they broke and I headed for the door, I shouted back, "No, Father. I told you already, I'm a Boland."

It was all of six months before I next went to confession, and then it was to a "travelling priest" from Mount Argus who came around our street a few times each year collecting money. He heard my confession in our parlour at my Ma's insistence, laughed his head off, and gave me three Hail Marys as penance.

I remember all kinds of films for all kinds of reasons, but the one which gave me mixed feelings of both pain and pleasure was *I Wonder Who's Kissing Her Now*, starring June Haver. The picture was shown in the De Luxe Cinema in Camden Street. I was going through Newmarket on the way there with a friend, when all of a sudden he ups and throws a big stone over a wall into a yard, where it hit a child who started screaming. After the stupid fool threw the stone, we both had no option but to run like hell. A few hundred yards and several streets on, we stopped running, thinking we were safe. But just then, a tall man, built like Tarzan, caught up with us from behind on a bicycle, jumped off it and came towards us. My friend, recognising him as the man from the yard, took off running without warning me. I stopped and faced the man, thinking he was looking for directions to some place, and not connecting my friend's take-off with the man's landing. Out of the corner of my eye, very shortly to be forcibly closed, I caught a glimpse of my friend disappearing into the sunset.

"Tarzan" was evidently the father of the child struck by the stone, and his first full-frontal punch on my face sent me staggering back against the wall and onto the ground, with blood spurting out all over my clothes and splashing onto the wall. Before I could get up off the ground, he kicked me in the ribs, the face, and the back. By this stage I realised that he was not looking for directions, that he knew his way around very well. As the crowds gathered for the execution, I managed to get upright and launch my puny teenage body at the enraged giant, for by this stage my legs did not possess the energy or speed to take me out of his running or cycling range. After landing a few weak blows to his face, I was glad to see that his face was also covered in blood, but unfortunately it turned out to be my blood.

After being punched and kicked several times more around the street, and almost on the point of unconsciousness, a few men pulled the monster off me, and I remember clearly one of them saying, "No matter what he's done, he doesn't deserve that." They helped me to my feet, and standing there swaying in the middle of the crowd, I took stock. One rapidly closing eye, one pair of lips like Paul Robeson, cauliflower ears that could have fed five thousand, bruised or broken ribs for lights out, and a new red jumper and streaked-red trousers. After they cooled the monster down, I dripped my way over to him, and Nancy Griffith-like "from a distance" said, "Mister, I didn't throw that stone, and I'm sorry your child got hurt, but some day I'll beat the shit out of you for doing this to me." While I was saying all of this, my blubbery lips never moved. The anger stayed with me for a very long time, but the more bloodbath fights I saw as I was growing up, the more sickened I became with the whole fighting scene. Anyway, the man was so big, I would have needed a sledgehammer to do the job on him.

Instead of going home or to hospital for treatment after the fight, I went on to the De Luxe to see the picture. But right through the picture my head throbbed, my lips were swelling

50

to the size of boxing gloves, and my good eye was slowly beginning to close. But I did manage to get a good half-eyeful of the beautiful June Haver, and my mind moved from the painful to what the priest would call entertaining thoughts. But for thoughts that really entertained, Rita Hayworth dancing in *Gilda*, or Marilyn Monroe doing anything, took a lot of beating.

When Errol Flynn was Robin Hood, we all made bows and arrows, and fought pitched battles in our farmyard. We all wanted to be the "chap" — the leading man — in our plays, and there were big arguments to see who would play Robin, Friar Tuck, Little John, and the rest of his friends. But everyone eventually got to play Robin, and even the Sheriff of Nottingham, although that was the least popular part. Maid Marian was played by little Lena O'Connell from next door. We lifted Lena up onto two bales of straw, and as the Sheriff of Nottingham fought to keep her prisoner, Robin fought to rescue her. Little Lena was about three-years-old at the time, and was constantly in and out of our house and yard. She followed me around everywhere, usually with a snotty nose, her thumb stuck in her mouth, and her other hand either holding up her perpetually half-mast knickers, or scratching her bum. She hardly ever said a word. Once in the middle of a battle, the Sheriff of Nottingham pushed her out of the straw shed to Robin Hood, saying that she had wet her knickers and he could have her.

The O'Connells moved away to England when Lena was about seven, and about ten years after that she and her family came back to visit. I just couldn't believe it was Lena. She had grown into a beautiful, sophisticated, articulate young lady, with not a knicker in the world to be seen.

When Errol Flynn was the boxer Jim Corbett in *Gentleman Jim*, we all became boxers and knocked hell out of each other. But enthused by the skill of the "gentleman," one or two of us went just that little bit too far and challenged Snotty O'Doherty of Chamber Street to a fight. Unfortunately though, Snotty had also seen the picture, and beat the shit out of us. Errol then led his troops to death and glory in *They Died With Their Boots On*, while poor old Olivia De Havilland had to stay home brokenhearted. For this picture we made wooden swords and charged up and down the street roaring "Charge," and whistling, humming, or la-la-ing "Garryowen and Glory," the music played throughout the film. I was able to make the battle scene a little more realistic, for I got out my pony "Gooseberry," and led the charge down Donnelly's Lane.

I was Hopalong Cassidy when we played cowboys around our yard, while there were several Roy Rogers', but only one Dale Evans — Lena. Gene Autry also made an appearance from time to time. Once when acting as the Lone Ranger, I was on the roof of one of the sheds in our yard shouting, "Hey-Ho Silver!" when the Da appeared out of the harness room with the long trap whip in one hand, and a ladder in the other. He quickly climbed up onto the roof at the end of the row of sheds, and came running down towards me with the whip flashing. I jumped down into the dung heap, intending to gallop down the yard and out through the gate to Cork Street and safety. As I jumped , I heard a cry from the Da, and looked back to find that he had disappeared. He had fallen through part of his beloved corrugated iron roof, and landed down in one of the pig sties. The Ma came running out, and I followed her to the pig sty to find him badly bruised and winded, lying in a pig trough with half a dozen pigs snorting at him, thinking perhaps that he was an extra dinner treat.

Although the Ma called me to come over to help him, I felt it might be unwise to do so, for he still held the whip firmly in his white-knuckled hand, and was glaring at me in a very unfriendly way, muttering something about "The yalla cur." As the Ma struggled to get him up on his feet, I said, in all innocence, from the safety of the outside of the half-gate of the sty, "If you'd been able to run a few feet further, Da, you'd have fallen into the hayshed and had a soft landing." This was too much for him. He was on his feet like a shot and with whip flashing, he chased me down the yard, as I fled to the safety of Cork Street, where I could out-run him. Many, many times I was run out of the yard onto Cork Street by the Da, and each time it took several days to cool him down. There were times when I ate my meals alone in the harness room, the workshop, or anywhere else it was safe to do so, until the cooling-off period had passed.

The other pictures I remember were *Captain Blood*, *The Black*

Swan, and all the "Hopalong" and "Charlie Chan" ones. The Da also liked going to the pictures, his favourite cinema being the Corinthian at O'Connell Bridge, nicknamed "The Ranch" because of all the cowboy films it showed. He would go to the Corinthian on his way back from paying his union dues in Liberty Hall. When he would say to me, "I'm thinking of going down to the union," I knew that the pictures and a bag of sweets were coming up. It was lovely going to the pictures with the Da at night, where I could sit on the soft expensive seats, away from the shouting, smelly kids of the locals for a change. After the pictures and the sweets, we would stroll hand-in-hand back over O'Connell Bridge, stopping to listen to the hurdy-gurdy man and giving him a penny. After that we would get the 50 bus home from D'Olier Street. In those days you didn't queue for buses, you just made a dash for your one when it came along.

The first Frankenstein picture I saw was in the Corinthian with my Dad. If ever a picture frightened me, it was that one, but I enjoyed it nevertheless, and was still enjoying it in a terrified sort of way, when the Da suddenly jumped up and dragged me from the cinema by the hand, saying, "This is too frightening for you." The truth was that when the part came where the monster smashed the chains that bound him, and reached down to pick up the little girl, it was too much for the Da, who, not to admit that he was frightened, rushed me out saying it was for my benefit. The hurdy-gurdy man got no pennies that night, for we crossed O'Connell Bridge like a rocket, in the Da's anxiety to get home to the safety of his own house. But more terror was to come for the Da that night, for after putting me on to the platform of the bus, he missed his footing as he jumped on, and was dragged from D'Olier Street, through Westmoreland Street, and around into Dame Street, before the conductor managed to get the bus stopped. His knees and shins were all cut and scraped, but after getting home and washing and bandaging them, he felt a lot better.

Pictures in the Rialto and Leinster cinemas ran for Monday, Tuesday, and Wednesday, with a new programme running the next three days. There was a separate programme for Sundays. Admission to the cinemas for children on their own was four old pence. The number of seats available for children was determined by the likely adult support for the picture being shown, and the size of the children's queue. For three consecutive nights I tried and failed to get through the children's queue to see *Robin Hood*. In desperation I asked an adult to get me a ticket for which I gave him the money, and that way I was able to go in and sit beside him in the more expensive seats, all for four pence. This was something a lot of kids did when all the children's seats were sold out. Kids couldn't do that these days.

Whereas most parents, despite their difficulties, seemed to be able to raise the four pence to give their kids a treat at the pictures on Sunday afternoons, very few of them could manage to raise the money to send them mid-week as well. I managed to go nearly every Sunday and usually one other time during the week. When there was a particular film on during the week which we wanted to see and there was no money for it, we collected jam jars from house to house and sold them to raise the money. Some cinemas, like the Lyric in James's Street, would take the jam jars instead of the money. I didn't go to the Lyric too often, but for a time, for a few weeks, I went every week to see a follyinupper named *Don Winslow of the Coastguard*. We considered the Lyric and Tivoli cinemas to be flea pits, but I dare say that the people from James's Street and Francis Street thought the same about the Leinster and Rialto.

The seats in the Lyric were long wooden benches like you would see in a park, but about five times the length. If there was a big crowd, the usher would just shout, "Move up there," and push more kids into the seats. We were so tightly packed at times that, if you got up to go to the toilet, you couldn't get back into your place.

So, many kids, rather than take a chance on losing their seats, would simply piddle on the floor just under their seats. About an hour into the show, a smell would rise up because of all the piddling done on the floor. And then the usher would come around with a disinfectant spray, and spray all of us, right over our heads — clothes, everything. I cannot remember anyone complaining, for usually we went out smelling a lot better than when we came in. Once when I was in the Lyric, a little fellow in the corner next to the wall dropped his pants and did a jobby. When the smell reached the nostrils of those around him, there were cries of, "Jasus, what's that?" followed by a mass exodus from the vicinity. But a few brave kids went and stuck the offending kid's nose in the jobby, and the brown-faced kid, shouting and crying, ran out of the cinema.

In those days, the pictures were shown in what was called "continuous performance" — first the short film, then trailers, advertisements and a newsreel, then the main feature, then the short again, followed by the rest of the programme, and they might even show the whole lot for a third time. This meant that people would come in and sit down in the middle of any part of the show, watch it all the way around to the point they came in at, and get up and leave. Because of this, the house would never be cleared between showings, and if you could avoid the usher's eye, you might manage to see the whole performance two or even three times. One drawback of this system was that people were always hopping in and out of seats, blocking the view and cursing and stumbling as they tried to find their way in the dark. It caused my Ma a problem on one of the few times I brought her to the pictures.

The Ma very seldom went out anywhere, preferring to stay in knitting or reading. But on one memorable occasion she came with me to see Errol Flynn in *They Died with Their Boots On*. We went to the front of the balcony in the Leinster for the early performance. When it was time to go home — we'd caught up to the part of the film at which we came in — the Ma left her seat like a flash and hared up the aisle steps towards the exit. But instead of turning right for the exit at the top of the first flight of steps, she kept on going, right into the gents' toilet, an easy enough mistake to make in the dark in a strange cinema. I was too far back to stop her, so I did the clever thing and went straight out into the street to wait for her, believing that someone would sooner or later throw her out, if only for molesting the men in the toilet. After a few minutes she arrived out, furious, and with a big red angry face.

"Wait till I get you home, you little bastard," were her first words to me, and for a genteel lady who didn't usually swear, that was quite a mouthful. But before we got home we made up. I promised not to tell the Da that she had broken into the gents' toilet, or that she was giving him rabbit in his sandwiches and telling him it was chicken, and she promised not to beat the lard out of me.

Not everyone could get to see all the pictures around in those times before television arrived, and so, if one kid saw a picture that none of the others had seen, he would be booked to tell that picture. As storytelling in one form or another was very popular at the time, some great picture-telling took place. All of the gang were good picture tellers, but some were more long-winded than others, trying to add in the music and even the sound effects.

One would say something like, "The picture starts with the coach going like blazes, with ten Indians after it, diddle-ah, diddle-ah, diddle-ah [the sounds of the horses' hooves], and the girl inside screaming for help 'Save me, save me.' Then the chap appears from behind a rock on a white horse, chasing after the Indians, shooting them off their horses one by one, bang, bang, budum, budum. And then he killed all the Indians and the girl jumped out of the coach into his arms just before it went over a cliff. Da, da, da-da, da, da, da-da, [the Wedding March]."

Sometimes the storyteller would run around, or jump onto the floor just to show how the baddie fell. But sometimes there would be shouts from the gang of, "Will you get on with the shaggin' picture and forget about the da-das and the dum-dums."

If in those days we had been told that some day a thing called television would come into our homes, where for hour after hour, every day, we could see all the pictures we wanted, we would have all thought that all our birthdays had come at once. Years later, when grown up, I returned for a holiday from working in England, to find that the Ma and the Da had a television set and were happily watching a very snowy-screened RTE programme. Shortly after that my Dad died. Television had come too late for him to get much enjoyment from watching his beloved cowboy pictures.

Some of my School Days

My days at the national school were the unhappiest days of my life. Within half an hour of starting in Weavers' Square National School, at the age of six, I did everything in my power to escape from the system. It was to take me six long years. My pals of the same age knew when they would be starting school, and some were looking forward to it, a bit like a duck looking forward to orange sauce. There was no mention at home as to when I would be starting, my mother hating the day when I would leave her. But after I begged my mother to let me go, a day or two after my pals started, she gave in and took me around to the school.

The door into Weavers' Square School had steel bars, through which mothers handed in flasks and food parcels to their offspring at break time. School had already started when we arrived, and after my mother filled in some forms, we were brought along to a classroom. I remember a tough looking woman teacher with a boxer's nose coming out to the corridor to speak to my mother, and then the three of us going into the classroom, to stand just inside the door.

"Does anyone know this boy?" asked the teacher, and a few little hands shot up in the air.

It all seemed nice and friendly, despite the look of the teacher. My Ma gave me a kiss, and with tears streaming down her cheeks, left me to start one of life's great adventures. First impressions that day looked good — that's why I never trusted teachers for the rest of my days at school.

The seats in the classroom were long benches without back supports, not the double or single seats that kids sit in these days. No sooner had the door closed after my mother, than the shouting started. The rest of the boys who had started a few days earlier had not prepared me for this. They had been given some lessons to learn, and were now expected to prove to the teacher that they had done so. For every wrong answer given, and there were quite a few, the teacher would roar and shout and bang her cane down hard on the desk with a cracking noise, making the child cry, cower, or put his hands protectively up over his head. Being a new boy, and not knowing any of the work, I was excused from this treatment until I was expected to know some answers. The first ten minutes or so in that school, did not impress me, but being a trier, I decided to give it a little more time before making my mind up about going or staying. Time-up for me, as far as school was concerned, came about twenty minutes later.

I was sitting beside a little fellow who was shaking with fear as the teacher made her way along towards him, asking questions, getting wrong answers, and shouting and banging the desk as she came nearer. When the little fellow gave the teacher a wrong answer, she shouted even louder, and banged her cane down hard beside his hands. The little fellow then did two things simultaneously, which I hadn't seen before, nor have I seen since. He started to cry, and shit down his legs. It may have been an occasion for those more experienced to take notes and contact the publishers of *The Guinness Book of Records*, but my immediate reaction was to put distance between me and the little fellow, which luckily I could do as I was sitting at the end of the bench.

"Where do you think you're going?" shouted the teacher at me, as I moved smartly across the room out of Smelly's way. Although only six, I was street-wise for my age, and knew what I liked and didn't like, and shit was something I definitely did not need to see or smell.

"I don't like to smell shit," said I honestly.

"Get back there, I can see that we're going to have trouble with you," said she, giving me a switch of the cane, but not imagining the extent of the trouble there was to come. Reluctantly I placed one cheek of my arse on the edge of the seat, keeping away as far as possible from the poor terrified smelly kid.

"Move in," commanded the teacher. I neither moved nor spoke. For a few moments we stared at each other, before she backed off. Maybe she saw something in my eyes which gave her a clue that I was fast becoming disenchanted with this school thing.

The teacher was obviously well prepared for incidents such as this which literally scared the crap out of the kids, for she produced some old newspapers which she handed to the little fellow, telling him to go out to the toilet and clean himself up.

As he was hobbling bow-legged out the door, the teacher handed me some more of the old newspapers and said, "Here you, clean up the floor, and go out after him to the toilet and clean him up too."

That was it for me. I had tried this school thing, I didn't like it, and I was going home. I walked out of the room, down the corridor, across the deserted school yard, and headed for the door with the steel bars, while the teacher ran after me shouting, "Come back here."

When I got to the door which led to the street and freedom, it was locked. While I was trying to figure out a way to go through it or get it off its hinges, the teacher arrived, shouting for me to get back to the classroom.

When I shouted "No," she pulled me by the arm, but let go quickly when I made a kick at her, something I would never have done in a street fight. Then the head nun appeared, and sent the teacher away while she reasoned with me. I still refused to go back, looking away from her, trying to figure out how I was going to dismantle the door, or find a way up onto the shed and down over the wall. When the nun's sweet-talking failed, she called in the heavy artillery in the shape of a teacher who was bigger and even uglier than the first one, the female equivalent of the Frankenstein monster. With a twist of my ear, and a bending of my arm right up under my shoulder blade, I was not so much marched as pushed back, all bent up, to the classroom. I did however manage to give the monster a hefty kick on the shin before being overcome. I had no qualms at kicking out, for it was obvious to me after only half an hour at school, that I had entered a vicious and vindictive world, where survival was the name of the game.

I was thrown back into the classroom, where I tripped, fell onto the desk, and rolled off onto the floor trapping my arm behind my back. I remember being hurt and in pain, but never a tear was shed, either then, or at any other time during my schooldays. But little did I know then, that compared to the beatings to come at the hands of the Christian Brothers, this arm-twisting incident was nothing. The shitty kid had by then come back from the toilet, and although looking a bit cleaner, he still smelled terrible. When instructed again by the teacher to sit beside him, I again refused. When handed the newspapers for a second time and told again to clean up the floor, I refused again. It was obvious that my first protest, my effort to escape, meant nothing. So I again headed for the door, the corridor, the yard, the outer gate beyond and freedom.

Having got to the gate and discovered again that I was unable to open it, I stood with my back to it, as again Frankenstein and my teacher approached, this time with canes in hand. The mistake that my teacher made when making a swipe at me with the cane was that she allowed it to slip out of her hand, and quick as a flash I grabbed it up. Holding the cane by its tip, I swung the crooked handle part over my head at the two teachers, who first stopped, then backed off slowly, then a little quicker, before suddenly making a dash for the school building with me after them shouting and swishing the cane. Just as I was about to catch them before they entered the school, the head nun appeared and asked them both to go in to her office. After a few minutes, as I stood out in the yard on my own, the head nun re-appeared and sweet-talked me into giving up the cane. After that, a reconciliation of sorts took place between me and the class teacher, the one with the newspapers. A new seat was found for me at the back of the classroom, where, in the circumstances, I was reasonably happy to stay until break-time, when my Mammy, who would be coming around with something for me to eat, could get me out of the God-forsaken place. But it was not the Ma who came around with my lunch, but Mrs McCann, a neighbour, who said that the Ma was sick. I told Mrs McCann through the bars that I had to get out, to go home to get the Ma messages. She told me that she had got them.

I did not return to school until about a week later, and then only after my Dad had persuaded me to do so. From then on, until I finally left school, no real effort was ever made to understand or help me, only to humiliate and beat me. In my second year in Weavers' Square school, I moved into Miss Regan's class, and sad to say, my one memory of that experience which lasted a year, is the day, when in a fit of temper, she broke the frame of a picture on the wall, so as to get something stronger than a cane to beat me with. My crime was that in answer to a question, I told her that the name of the hair which grows on a horse's neck was the mange. The answer of course was the mane, but so nervous was I that I blurted out the wrong word.

My next school was the Christian Brothers, and although I had not done terribly well in Weavers' Square, I was looking forward to my new school and hoping to do well there. But that was not to be, for very quickly I encountered Brother "Killer" Kelly and Mr "Manky" Monahan. There were rumours going around that they had taught the Germans and the Japanese the best way of torturing people. My big problem was the Irish language, learning it and doing other subjects through it. My mother, being from Fermanagh, had not learned it at school. My Dad was not just negative about the language, he was positively hostile to it. Although a decent and fair man generally, he had a thing about the Irish language and some of those who spoke it.

He would say, "Them bloody country people shouldn't be allowed in to Dublin at all. If they're not talking Irish, they're wearing Pioneer pins, playing Croke Park football, starting up Legion of Mary clubs, or forming little cliques of one kind or other."

The Da would say things like this, more in fun to take a rise out of the Ma, rather than to offend those born beyond the Pale. The Ma's response was usually swift and cutting, reminding him that it was the treacherous Dubliners who had gone on to the streets in 1916 to call for the execution of the leaders of the Rising, that it was one of them who had shot Michael Collins in the back, and that if she had her life to live over again, she wouldn't put her foot in the bloody dung heap of a city. By this stage my Dad would be in stitches of laughter, while my Ma would have a big red angry face.

In order to rub it in more, she would walk around the house after the Da singing,

> "Who fears to speak of Ninety-Eight,
> Who blushes at the name,
> When cowards mock the patriot faith,

And hang their heads in shame …"
After giving the Da the whole fourteen verses of *Ninety-Eight*, she would then give him another ten or so verses of *The Bold Fenian Men*, followed by *A Nation Once Again*, and for good measure, she'd throw in Robert Emmet's speech from the dock. If he was lucky, a short fiery tirade against the Dubliners would follow:

"If it wasn't for the country people who came to live in this dung heap of a place, it would be full of English lackeys, for the Dubliners are a gutless lot."

After these encounters, which fortunately did not take place too often, there would be an atmosphere in the house for a few days, during which dinners would be more thrown at the Da than placed before him. And the odd day she would "forget" to pack his lunch. These explosions by the Ma may have been entertaining for the Da when he started them off, but the consequences were not too happy for him, for as proud as my Dad was of being a Dubliner, the Ma was equally proud of not being one. Usually though, after a week or so, they would sit and laugh about the exchanges, she calling him a terrible man for having annoyed her, and he calling her a touchy country woman for not being able to take a joke.

The teaching philosophy of the Christian Brothers at the time was to beat the knowledge into you. And although there are times in all our lives when we can look back philosophically on hurtful or unpleasant experiences without rancour, it positively makes me sick to hear grown men who have suffered at the hands of the Christian Brothers saying, "They were marvellous teachers. Look where I got today because of them."

I believe that for everyone who came through their system and "got" somewhere, there were thousands whose lives were destroyed, both emotionally and career-wise. Although there were obviously some Brothers who were kind, and reasonably good teachers, the majority of them, from my own experience and the horror stories I have heard from others, were cruel, uncaring, and poor teachers.

Any outside chance I had of mastering the Irish language soon disappeared, when as a result of pulling a jennet's tail, I got a kick in the chest which sent me out through a stable door, and on to hospital, and then to bed for about three months.

After that the Irish was lost forever, or at least as far as the national schooling system was concerned. There were days in school when only Irish would be spoken, including in the school yard. To enforce this, traps were set and teachers' pets used to spring them. This was done by the pets offering a toy or some kind of object to an unsuspecting kid, who might say, "What's that for?" That question, in English, was enough to have him reported to the teacher who would flog him — no better way to describe it — in front of the class. No wonder our beautiful language is almost dead, when those sadists who were entrusted with passing it on to the children of the nation, instead of doing so with the love, gentleness, and encouragement which such a great treasure deserves, did it in such an uncaring and brutal manner, that generations of us have been turned off the greatest part of our heritage.

"Killer" Kelly was a sadist if ever I saw one. Not content to use the heavy stitched leather strap on us as it was, he made it more deadly by splitting open the leather on one side and inserting three-penny pieces. This addition of metal caused not just pain, but also injury. I remember arriving home from school with my two hands swollen to the extent that I could not lift a knife or fork to eat my dinner. And that was quite apart from the pain in my ears where I had also been boxed. Thank God that teachers can no longer assault children in this manner, or, if they do, they can be prosecuted and shown up in court for the bullies they are. Whereas my mother would sometimes cry when I came home injured from the beatings, my Dad, so typical of a lot of parents at the time, would say that I probably deserved it. And we children accepted the beatings as being just part of a normal schoolday. But thank God those days are gone, for today's teachers are highly trained caring people, seeing their work as being vocational and challenging, at the same time unique and important, in that they are entrusted with the nation's greatest treasure — its children. For this reason alone, teachers are one category of workers who deserve the very best of working conditions and support.

"Killer" Kelly was a dead-shot with the strap. If a fly landed on the door or wall, he could throw the strap and squash it from about ten paces. Indeed with this prowess, coupled with his mental make-up, he would have done well in the army, or a slaughter house, had he not decided to follow the religious life. Unfortunately we were easier targets than flies. But there were two days that I remember when the hot-shot Kelly got things wrong.

Once, when he was beating a boy with the strap, the boy took his hand away, and Kelly struck a pocket in his cassock, somehow igniting a box of matches. Nothing happened at first, apart from Kelly losing the head altogether with the boy. But after a few minutes, we were amazed and delighted to see smoke pouring from Kelly's clothes — he was going up in smoke before our very eyes! It was as if all our prayers were being answered at the same time. One minute he was just standing there, tongue clenched between his teeth as he dished out punishment, and the next he was jumping around the place, shouting and trying to get his "frock" off as the smoke poured from him. Unfortunately he survived the fire … As a kid said at the time, what we needed was some sand to pour over him, about three tons of it.

The second time Kelly's marksmanship with the strap let him down provided me with another special memory. It was one of those hot, hazy, summer afternoons coming up to holiday time, when, due perhaps to the stuffy air and for once a quietish atmosphere in the room, some of us were dozing. Kelly spied me at the back of the class in slumber mode, and slowly aiming the strap so as to bring about an awakening, he let fly. Unfortunately for him, however, but more so for the boy in front of me, he missed me and hit him a terrible smack on the face. The injured boy, a quiet kid called Paddy Doyle, let out an unmerciful roar, which brought a worried-looking Kelly running down to him, more out of concern for the repercussions to himself, I would say, than for the boy's injuries. As Paddy held his hands up to his face, and the blood ran out from between his fingers, he let loose with a string of curses at Kelly, which surprised us, and at the same time qualified him for membership of our gang. He pushed Kelly away and headed for the door, shouting that he would bring his Daddy back to him. Many of us, after beatings, said that we would bring our Daddies around to take care of Kelly and Monahan in particular, but none of them ever appeared — until that day, that is.

In the excitement of the moment we had all come awake,

alert as we had never been before. Kelly left the room for a few minutes and returned composed and quiet for a change. We were working quietly at our desks when we heard shouting, then the banging of a door, as the noise came nearer to us. Then suddenly the classroom door burst open with a crash and a splintering of wood, and there stood a very big man with a little boy by his side wearing a blood-stained bandage around his head.

"There, Daddy, that's him, he did it!" shouted Paddy, pointing at Kelly.

Like a flash, the Daddy was across the room, grabbing Kelly by the throat with his left hand, and giving him a belt with his right.

"Come on, Mr Doyle!" we all shouted. But Mr Doyle only managed to get a few belts at him before other teachers came running in and pulled him off.

But those few belts did the trick, for "Killer" Kelly lay on the floor, having got a dose of his own medicine. After that, Kelly was missing for a few weeks, and Mr Kerrigan, a decent kind of man who didn't use the strap too often, took control. We were all hoping that "Manky" Monahan wouldn't be put in charge of us, for he was almost as bad as "Killer" Kelly. Monahan was a countryman, and, because of his accent, we all had difficulty understanding him. Someone said that he was inside the G.P.O. in 1916, and that was why he got his job as a teacher. We all agreed that it was a pity he ever got out of the G.P.O.

Paddy Doyle returned about a week later to a hero's welcome. After that the teachers left him alone, and I mean alone, not taking the slightest bit of interest in him or his work.

We said prayers at the start and finish of each day, one of them being for the soul of Brother Ignatius Rice, the founder of the Christian Brother movement. The prayer some of us said was not the one printed in the book, and if ever the prayer was answered, the only way he would ever be canonised would be out of the barrel of a gun.

In Kelly's class I was beaten for the usual reasons, but also for heading a goal in a Gaelic football match.

During the war when the Germans bombed the North Strand they also dropped one on the South Circular Road, in the vicinity of our school. Ever after that, I said a special little prayer each night, "And please God, let the Germans bomb that bloody school of ours".

For part of my last year I had a teacher who didn't rely too heavily on the strap to keep order, and I began to work better. But unfortunately he got to me too late, for the beatings and the sarcasm of the teachers who had gone before him had already done the damage. I remember a Christian Brother pointing a round thick stick at me and saying, "Boland, you're a fool. What are you?"

When I hesitated to answer, he came menacingly towards me, but having had enough beatings that day, I quietly answered, "I'm a fool, Sir."

There were teachers along the way who wanted to see me cry, but they never did. I just put out my hand, stared them straight in the eyes, and took it all, from loaded straps to sticks, without batting an eyelid. When Kelly beat me, he would sometimes hold me in a bent position with my hand across his knee and his tongue stuck out, clenched between his teeth, as he laid into me viciously. There were times, looking at his face twisted up in anger, when I wished the bastard would bite his tongue off.

The happiest time in the classroom was once a week when we all stood around the side walls, singing. I liked to sing, and one happy memory is of standing with my back to the window as the sun shone in, making a shadow on the desk of my head with the sticking out ears, as we sang,

> "There's music in my heart all day,
> I hear it bright and early,
> it comes from fields far far away,
> the wind that shakes the barley."

It is sad that is the only happy memory I have of my school days with the Christian Brothers.

The final memory of my schooldays with the Christian Brothers was an incident which happened an hour or so before we broke up for our summer holidays. The teacher, in a playful kind of mood, was telling the clever boys in the class what jobs he thought they would work at in life. One by one, he pointed to various boys saying:

"You'll be a carpenter. You'll be a soldier, You'll be a baker,"

66

etc. Then he pointed at me and said, "Boland, you're a fool, you'll never be anything."

That was my last day in school with the Christian Brothers. At the age of twelve, having been humiliated by beatings and sarcasm from the age of six, my confidence was not just destroyed, but devastated. This "playful" humiliation was the last straw.

I had stayed away quite a bit from school, not by mitching, but by persuading, even begging my mother to keep me at home, because of some upcoming beating I knew I would get for not knowing my Irish. There were times however, when it was all too much for the Ma. My Dad never knew about all of my absences, although there was the odd time when he himself kept me at home to mind a sow, or to send me to work in his place, driving the horse and cart in the Corporation while he attended to some business matter. My mother was very soft with me, and I thank God that she was, because of the cruelty of school. There were occasions when the school inspector called to our house to see why I was absent. I think the Ma ran out of excuses, for once she had to engage a solicitor to keep me from being taken away to a reform school in Artane, or one in Glencree. The fear of going to Artane, and having to play in their bloody Artane Boys Band in Croke Park, was almost enough to make me want to go to school, but not quite. After I packed up school, there was still some trouble with the Department of Education, which tried to get me back into the system, but in due course, they too, like the Christian Brothers, gave up on me.

Once, before I left school, the Ma tried to get me to go on one particular day, and went so far as sending for the Da to come home from work to make me go. But I locked myself into my room with my little dog Flora (the massive Rover had not yet appeared on the scene). When eventually the Da broke open the door to get me, I had long gone, via the drainpipe, and he had missed a day's wages. I had to stay away from the Da for a few days until he quietened down.

Work and Back to School

That summer should have been a happy and relaxed time, as I was on school holidays, and didn't have to worry about "Killer" Kelly or "Manky" Monahan. But it was a tense, unhappy time, for at the beginning of the school holidays I had announced at home that I was finished with the Christian Brothers, having had enough of the beatings and the sarcasm. I was not sure that I was going to get away with my plan, but only time would tell. I told some of my pals about it, but they just laughed, saying I was a right eejit if I thought I could get away with it.

"They'll lag you off to Artane if you don't go to school, and then you'll have to play in that stupid band for all the culchies in Croke Park," was one of their comments.

Another one was, "They'll take you off to Glencree and lock you up for seven years, and when you get out you'll be a Christian Brother with a big strap, ready to beat the shit out of every kid in the school."

Another kid said that I'd be sent to a place in Kerry where they would make me into a priest. And if all of that wasn't enough to worry me, Birdbrain said they would send me to Cork and make me into a nun. The nun bit was the last straw, and I can tell you that marching around Croke Park wearing those ridiculous looking cloaks, and playing *A Nation Once Again* for the country people, was beginning to look better and better as the summer progressed.

A week or two before the back-to-school date, all hell broke loose when my Dad arrived home one day with a new schoolbag for me. That's when I made my last stand, when I pleaded, begged, cried, and finally threatened to run away, if sent back to the chamber of horrors. Friends, neighbours, anyone who could walk, hobble, or even be carried, were brought in to the house to try to persuade me that my schooldays were the happiest days of my life. If they had been beaten every day by

"Killer" Kelly, they'd have played a different tune on their fiddles. But after hours and even days of this shuttle diplomacy, I was eventually moved somewhat from my original viewpoint, "moved," that is, to the point that I was even more determined to give up school. My "last stand" started at 11 o'clock on a Saturday morning, and lasted through blood, sweat, and tears until 6 o'clock, when I collapsed from sheer exhaustion and hunger, not having been able to eat all day through the upset.

I was put to bed early on the Saturday, and awakened the next morning to go to Mass. The Ma and the Da were very nice to me, and I felt that they were at last coming around to my point of view. I was given a big breakfast, and when invited to go to Mass with Da, an honour not dished out every day of the week, I readily agreed.

But midway through the Mass my heart sank when I heard my Da's prayer, "Holy Mother of God, send Paddy back to school without any more rows."

With that I took off out of the chapel and went home to pack, for there was nothing left to do but leave home. The Da left Mass early too and met me at our hall door as I was dragging myself away from the Ma who was in tears. The Ma and Da then asked me to go back into the living-room where I sat opposite them on the couch, shivering and crying, such was my upset and my fear about going back to school.

As I fiddled with the packet of bread-and-butter sandwiches which I had made up for my trip to God knows where, the Da looked at the Ma and then said, "Don't worry son, if those hoors are as bad as you say they are, you won't have to go near them again." With that, the tears just gushed out, and I threw myself into their arms, hugging and kissing them, and telling them how much I loved them. That was a special moment of happiness in my life.

But the next moment my world was turned upside down again when the Ma said, "What school would you like to go to then?"

Then the Da said, "He's had enough of school, he is either going to work for me or serve his time to a trade, isn't that right son?"

I relaxed again, started to cry again, and hugged them again. I had made my last stand and won, unlike a certain American general who tried a similar feat on the plains beneath The Little Big Horn. If old General Custer had had a few Christian Brothers in his army, they'd have massacred the poor Indians. A couple of belts from a strap with three-penny pieces sewn into it, and they wouldn't have known their tomahawks from their wigwams.

To make the decision to leave school and be supported by the Ma and Da was one thing. To stay out of that stupid Artane Boys' Band, or not be sent off to become a Christian Brother or a priest, or fate of all fates, a nun, was something else. However, in time the Ma and the Da organised the school departure, and I had no fears on that account. For about three weeks after school reopened, I was particularly happy. For the first week I walked around to the school at starting time, teasing the lads about having to go. Once I looked through the school gate and saw "Killer" Kelly, and for a little while we stood staring at each other, I the victorious one, as I thought. For a few days I continued to meet the lads coming out of school, and then on the last day of doing that, a great sadness came over me. I knew then that I had lost out on something great and wonderful, and all because of the lack of Christianity in the Christian Brother movement.

Across the road from where we lived, a motor body repair shop had just started up, and it was there that my Da got me a job to serve my time as a coachbuilder. I worked away happily for some time, but always there was this feeling in me, the feeling that things were not right. I started developing a desire so way-out, so mind-boggling, that it even frightened me to the extent that, for a long time, I couldn't bring myself to mention it to anyone. Then one night, as the Ma, Da, Albert and me sat by the fire, listening to the radio and yawning — for Albert the cat was always yawning and making the rest of us yawn as well — and as Bing Crosby sang, *Where The Blue of the Night Meets the Gold of the Day*, I seized the moment and blurted out my secret to the Ma and Da.

"I want to go back to school, I think I've got some brains." They nearly fell off their chairs, and were dumbfounded for a few moments.

"You mean to say," said the Da, "that after all the feckin' trouble we've had to get you out of school, from stopping you having to play in that stupid band, that now you want to go back to those hoors of Christian Brothers?" Then he added, "Are you sure, do you think you have a few brains after all?"

"Don't be stupid," said the Ma, "he's got more brains in his little finger than you have in your whole body."

"I'm not going near the Christian Brothers, I'm not that stupid," said I.

"What school would take you now?" asked the Da.

But I had it all figured out, and told them my plan. I was to give up my job, which saw me doing very little real training work — just cleaning the yard and going on messages. I had saved up some money and could afford to buy a few store pigs, if the Da would let me keep them in the yard. Then if he let me use the horse and cart, I could gather slop from door to door in the mornings to feed them. That way, after the sale of the first batch of pigs, I would be in a position to buy more, and be able to pay my way through a private school in the afternoons.

The Da was not too enthusiastic about the whole idea, saying that I was throwing away a good trade. He then asked if I would like to work in the Corporation, making tea for the workmen on the street, and later maybe getting to work a pick and shovel. The Ma, drooling all over me with tears in her eyes, told the Da that I should go back to school, that one day her dreams of seeing me with a briefcase and a cheque book would all come true. The thought of seeing one of his offspring so adorned was too much for the Da, and he agreed that a briefcase and a cheque book would be grand things for me to have. He conceded that if going back to school was the way to get these status symbols, then I should go, and of course I could get the pigs and borrow the horse and cart.

The next day the Da had an even better idea. He would buy more pigs, in fact get into the pig business in a big way. I would collect the feed and look after the pigs, and he would give me pocket money each week and pay my way through private school. It was the deal to end all deals, and in the traditional way that he sealed a deal with others, he spat a big gollier onto his toil-worn hand, and slapped it down on my scholarly hand, soon to be signing cheques and carrying a briefcase.

Just at that time the Da had a stroke of luck, because Jack Murphy, a friend of his who did the announcing at the horse trotting races at Raheny, had been offered the contract of removing the offal from the Clarence Hotel. Learning that the Da was getting into the pig business in a serious way, and not having pigs of his own, he offered it to him, the arrangement being that Jack's workman would collect from the Clarence each day, and Da — me in this case — would collect it from Jack's place in Kimmage. Thereafter, every day for years, I drove the horse and cart up to Jack's place on the Kimmage Road, to load up this fabulous pig swill. At that time the many small pig yards around Dublin either did not use, or have available, the effective pig feeds that are on the market today. It's amazing what you can learn, and indeed earn from the offal of a posh hotel, for that was how the Clarence was then regarded.

The swill, or slop, if you prefer, collected from door to door around Kimmage, Crumlin, and our own neighbourhood, consisted mostly of potato skins, cabbage leaves too badly withered to cook, and other odds and ends which just could not be eaten. This reflected the poverty of the times. It was said that in Rathgar, where because of the address, people spent most of their money on clothes to keep up an appearance, there was no decent pig swill, only tea leaves. But the swill from the Clarence Hotel was totally different. The 45-gallon drums into which it was loaded for taking away were chockablock with first-class cooked left-overs, the dry waste such as uncooked cabbage leaves and potato skins forming only a small part of it. This swill was ideal for mixing in with the poorer quality door-to-door swill. Both kinds of swill, together with grain drawn from Guinness's and Danny Norton's place, were mixed together and boiled in vats which were specially made for the job. This was powerful feed, and you could almost see the pigs growing. If you couldn't see them, you could certainly hear them, and there was no doubt at all about being able to smell them. But I tell you, anyone who got a rasher, or a Donnelly skinless sausage from any of our pigs, could taste that little something extra, thanks in some degree to the chefs of the Clarence Hotel.

The swill from the Clarence Hotel, or any other hotel for that matter, could tell you a lot about how the other half lived, for it contained the remains of cooked turkeys with only the breast, or just the legs removed. Chickens were similarly discarded, as were massive steaks with only a forkful taken off, roast beef in big chunks, ribs of beef, big lumps of ham and corned beef, as well as whole chops, and heaps of sausages and rashers. I can tell you, our pigs lived better than some families in our neighbourhood.. Considering what some families barely survived on at the time, it had to be wrong that so much good quality food was wasted by those dining in the Clarence. And then there was the waste of silver cutlery. After accumulating about a dozen knives, forks, and spoons bearing the Clarence insignia, my mother sent me down to the hotel to return them to the chef. When I asked to see him, I was sent around to the back entrance, where a big dirty-looking ignoramus of a man took the parcel off me, looked in and saw what was in it, then grabbed me by the collar, and warned me never to show my face there again. Thereafter, needless to say, we had the most stylish cutlery of anyone in the street, the quantity of which grew and grew over time.

Another rich harvest given up by the Clarence was the meat bones, particularly the large beef rib bones. I discovered that if I filled up a sack of them, O'Keeffes the Knackers would pay me half-a-crown or three bob for them. I was on to a good thing with the bones until the Ma got wise and confiscated some of the proceeds to stretch the house-keeping money.

My mornings were busy, yoking up the horse and cart and going to Guinness's or Danny Norton in Watling Street to load up with grain. Then on down to John McGrath in John Street, to load up the swill that he collected for us in the area, and for which we paid him about three shillings a barrel. At that time we were paying Jack Murphy six shillings a barrel. After unloading in the yard, it was then on to Jack Murphy and my regulars in Kimmage, to collect another few barrels of swill and the Clarence offal, before heading back to the yard to check through it, to make sure there was no glass or other objects likely to injure one of the pigs. The swills were then mixed together, the grains added to it and boiled, before being fed to the pigs. It was murder getting into the pigsties with the buckets of feed, for the little devils would jump up on you, nearly knocking you down before you got as far as the troughs. After the pigs had eaten, they were allowed out into the yard for a bit of exercise, and to do what little pigs usually do after their dinner, and sometimes in the middle of their dinner, and even at times on top of their dinner. When they were out exercising, the sties would be cleaned out, hosed down, and new dry sawdust spread. But once they had eaten enough, and attended to the call of nature, they were as happy as pigs in chiffon.

Farmers returning from the vegetable markets each week with empty carts would call to our yard looking for manure to fertilise their crops. Sometimes we would sell it to them, but mostly we would barter it for hay or straw. My mother, who had always fed the smaller number of pigs we kept before getting into the business in a big way, helped me in the yard at times so that I could get off in time for school.

I remember one particular day when I was about fourteen. I was on my way up to Jack Murphy's with the horse and cart to pick up the swill, when the heavens opened and the rain just soaked me through. Normally in such weather, I would tether the horse to a lamp post and take shelter. But on this particular occasion I was running a bit late, and not wanting to miss anything in school, had no alternative but to press on. At that time, plastic coats had not been invented, and rubber capes, because of the rubber shortage caused by the war, were hard to get. So, like my Da, I wore a heavy woollen coat to protect my shoulders and part of the rest of me from the elements, and on my head I wore a sack made into a hood with rest of it trailing down my back for extra protection. On this day we had no spare barrels into which to load Murphy's swill, and so it would have to be carried loose in the cart. On the way home, with the rain falling relentlessly and soaking me right through, and the sloppy swill coming up over my ankles and going into my shoes, I sat in the middle of the cart feeling particularly miserable and sorry for myself, wondering what life had in store for me.

If at that moment someone had told me that one day I would be in charge of the European manufacturing operations of an American multi-national company, hold senior positions in accountancy and personnel, chair meetings at conferences, hold a company directorship, attend meetings in far-distant places, and be met with limousines at international airports, I wouldn't have believed it.

In the beginning of my new school life, I went to a private commercial college in Abbey Street, initially three days a week, and then every afternoon only. I was the youngest at school, the others all offspring of well-to-do parents, who, having left school following their Intermediate or Leaving Certificate exams, were now doing commercial courses. When the work built up and I had to change my school attendance to the afternoon, it was a big rush to get the farmyard work finished and be in school on time. One of the big problems in getting ready for school was changing out of my smelly work clothes, and trying to wash myself in the big wooden bath. We did not have a bathroom in our house, and so once a week, on a Saturday night, I had my big wash in the wooden bath in front of a roaring fire in the living room. On workdays, however, prior to going to school, the washes had to be quicker, not as far-reaching, and they were certainly uncomfortable. The Ma and Da took their weekly baths in the same way as me, before a big fire, but at different times. Once when the Ma opened the living room door after having her bath, and Albert the cat shot out, the Da said, "I told you not to take off your clothes in front of the cat, you'll scare the life out of him." A wet towel around the Da's face was her response.

No matter how much I washed, and no matter how clean the clothes I put on, particularly during the week when I especially wanted to be clean for school, I could never fully get rid of that peculiarly awful smell of pigs, swill, manure, and whatever other perfumes farmyards throw up. Having one day taken my seat in school for the first time between the girls, the beautifully perfumed, washed, and well-dressed students started moving away, leaving me in no time at all surrounded with rows of empty seats. Quite honestly I couldn't blame them, for I must have smelled like something the cat dragged in a year earlier and forgot to take out. The college in Abbey Street was close to the Liffey, and at low tide on hot summer days there was sometimes a strong stench from it. I was not very comfortable sitting in a room with girls at the best of times, and particularly so on one very hot day, when one of them remarked that the smell from the Liffey was very bad that day.

Then a female voice with a grand Rathgar accent piped up

from somewhere behind me, "That's not the Liffey, it's your man from Cork Street."

Mr Mullery the baker took an interest in my new-found quest for knowledge, and made a significant contribution by sussing out another private school which had just opened in Dolphin's Barn, right on my doorstep. Transferring to that school was the best thing that happened to me. The two teachers who ran the place were encouraging, courteous, and nice in every way. In fact, looking back, I realise that they had created the perfect environment for getting the best out of children. The classes were small and mixed, but the children never once complained about the smell from me, maybe because, as one of them remarked years later, they too were part of the great unwashed, and if they got a funny smell, they might have thought it was from themselves. I thrived on the work, hungering for knowledge in this new learning environment. I became a great reader, and never wasted a minute at school. When after about a year, I won first prize, getting full marks in a series of exams, I was over the moon with delight and pride. When remarks such as "Excellent," "Well done," "Keep up the good work," started appearing all over my school books, I walked taller than John Wayne or Val Doonican ever thought possible. For years I went to this lovely school of encouragement and confidence-building, where the teachers never once spoke discourteously or raised their voices to the children, a far cry from the beatings and sarcasm I had experienced at the hands of the Christian Brothers.

While all this education was being laid on me, pigs were being bought, sold, fed, cleaned, swill gathered, horses shod, and the stables, sties, and yard cleaned and maintained. But at the age of seventeen, my life took a new turn when I got my first commercial job with Irish Press Newspapers. There, for a few years I spent the happiest working years of my life, in a lovely friendly atmosphere, amongst some of the nicest and most honourable people I have known. But the studying still went on and on. The start of my job saw the finish of the pig business, for the Da wanted to take things a bit easier, and after all, the business had accomplished what it was intended to do.

Taking the Pigs to Donnelly's

Selling pigs to Donnelly's bacon factory was a big business for us. Every few weeks we would send in a batch of "finished" pigs, meaning that they were the ideal weight for slaughtering, and would therefore qualify for the top market price. We got the pigs in the first place by either breeding them in our own yard, or buying them in from the market as "stores," half-grown pigs, ready for fattening. The feeding we used, as explained elsewhere, was first-class, and my father maintained that this, as well as the expert care we provided, made them the best pigs going into Donnelly's.

The sties in which the pigs were housed were purpose-built, and in this respect the Da was ahead of his time in designing and building them. They were laid out on two levels, one for sleeping, and the other for feeding and doing other things on. Each day the sties were cleaned out, hosed down and rebedded afresh. The pigs were hosed down daily, and were even scrubbed at various intervals. Indeed, if there was a prize for the condition and appearance of pigs entering Donnelly's bacon factory, ours would have won it.

When pigs were nearing their finished weight, getting near the time to be sent to Donnelly's, my father would ask Mr Williams from down the street to give us his opinion as to their readiness for the bacon factory. Mr Williams was one of the pig killers in Donnelly's, and what he didn't know about pigs was not worth knowing. Mr Williams would run around the sty after the pigs feeling their backs before announcing either, "They're ready," or "Give them another week." When they were ready for Donnelly's, we would then hold them until Mr Williams would tell us when to send them around, whenever Donnelly's, due to a shortage of pigs, would increase their prices. On the appointed day, Mr Williams would come around at his lunch time and help us to hunt them around to the factory.

The procedure was, when the sties were opened, the pigs would amble down the yard to the yard gate, one side of which would be open, and then they would enter Cork Street, and be shooed left by Mr Williams. About 40 yards further on up the street, they would be turned again into Brickfield Lane, or Donnelly's Lane, as we called it. Once in the Lane, the pigs had only about another 50 yards to travel, straight in through the gate of Donnelly's factory, and beyond to special pens to await their doom. Getting the pigs to Donnelly's was a simple enough exercise. Or so it seemed …

Mr Williams was a tall man. He seemed always to dress in dark trousers held up by a massive black belt, tied rather than buckled in the front. I was told that he sometimes took this belt to his kids. He usually wore a dark jacket and a striped open-neck collarless shirt. His skin was deathly pale, and he was serious and quiet-spoken. If ever a man seemed suited for his work, it was Mr Williams.

I have special memories of the Williamses, for they were a smashing family. Mrs Williams was a small, dumpy, red-faced woman who shuffled rather than walked. Mr and Mrs Williams lived with their five children in a two-roomed flat just down the street from us. They shared the house and an outside toilet with a family overhead. My memory of their flat is that there were beds everywhere, in both rooms. Mrs Williams' soup, with plenty of barley in it, was famous around the street, and now and then she would call me in for a bowl of it. Times were not easy for clothing and feeding a big family. In a street that had its fair share of characters, the Williamses, for all their struggles, stood out as a really nice family. And despite having quite a job to feed her own family, Mrs Williams could often be seen bringing some of her famous soup to families who had fallen on hard times.

In those days people in our street, as I am sure was also the case in every other street in the city, would never see a family go hungry. But such was the poor quality of the food, generally speaking, that it was necessary to augment it for health purposes, with some kind of laxative.

The most often used laxative in our street was castor oil, which had sometimes to be force-fed to a child. The usual method was to hold the child's nose while someone else poured down the oil. It tasted terrible. Another laxative I remember was a dark chocolate called Brooklax. There was a family living down the street who were so poor that they couldn't afford to buy laxatives, and it was said that the Daddy had to line the kids up on potties and tell them ghost stories.

One summer's evening after work, Mr Williams called in, felt the pigs, and announced that they were ready for Donnelly's, and as the price they were paying for pigs was right, they should be sent in the next day. My Dad arranged with Mr Williams to call down at lunch time the next day and help to hunt the pigs around to the factory. The next day, however, I decided to hunt the pigs around myself. After all, I had only to get them out the gate, turn them left, and 40 yards up, turn them left again, and it was then straight in through Donnelly's gate. Nothing could be simpler, I had done it umpteen times before with Mr Williams just standing there, doing nothing but saying, "Shoo, shoo," while I did all the real work.

First I opened one of the yard gates onto Cork Street. Then I opened the pig sties. The pigs, as usual, ambled out and down the yard, out onto the street. But instead of all of them turning left, as they were supposed to, and had always done, two little devils turned right, and took off down the street, one turning into the convent grounds, and the other heading for Ardee Street. While I chased around the convent grounds trying to get that pig out, another one went into Sheehan's shop and shit on the floor, and three others went into a nearby furniture factory, and scared the lives out of some of the girls who ran out onto the street screaming.

While all this was happening, the main body of pigs had passed by the turn for Donnelly's Lane, and was heading up Cork Street towards the junction with Marrowbone Lane and Donore Avenue. I managed to get the convent pig back up Cork Street, to join his three pals from the furniture factory, which by now had been driven back out onto the street, with the screaming girls back inside at work. When just at the turning into Donnelly's Lane, I looked back down the street to see that the Sheehan's shop pig had been hunted out and was now on its way down the street in the opposite direction.

I immediately abandoned the other four pigs and sprinted

down the street to round up the stray, when he too suddenly darted into the convent grounds. It took me ages to round him up and get him back on the street again. But instead of him turning left towards Donnelly's Lane and the other pigs, as he was supposed to do, he turned right, and right again, into Ormond Street. I eventually got him back into Cork Street, going in the right direction, as I chased him and the other pigs, which had by now passed by Donnelly's Lane.

By now I was in a panic, heart pounding and the sweat dripping off me, knowing all too well that if I lost any of these valuable animals, I needn't go home. As I dashed past Donnelly's Lane, feeling a bit like one of Frankie Lane's *Ghost Riders in the Sky*, I looked down the lane and saw two pigs just standing there staring at me. Just past Donnelly's Lane was the fever hospital, and there, grazing peacefully in the grounds, were two more pigs. Beyond that I came to Donore Avenue and the Marrowbone Lane junction where my worst fears were realised. To the left, up Donore Avenue, I could see more pigs, and straight ahead there were some more heading up Cork Street towards Dolphin's Barn. A few steps further on, I looked down Marrowbone Lane to see two more heading for the Maryland turning, while another two went straight on down the Lane towards the many turnings off it.

I ran up and down Donore Avenue, Marrowbone Lane, Maryland, Cork Street and Brown Street in an absolute panic, chasing the pigs as fast as I could, and shouting for someone to help me. But whereas the Dubliners of the area would feed, maybe even die for their own, none of them seemed to want to have anything to do with this mad kid chasing pigs up and down the street. Eventually, with a rush of relief which I can almost now experience reliving the incident, I got the pigs back to Donnelly's Lane, and in through the factory gate.

As Mr Williams was making out the receipt for them, and I lay slumped against the wall, exhausted but happy, he said to me, "Have you been running? You look very hot." Then he asked, "Had you any trouble getting them around?"

"No," said I, lying through my teeth.

"There you are now," said Mr Williams, handing me the receipt, "fourteen pigs received safe and sound. Did your father keep three back then?"

80

On hearing his question, I nearly dropped dead on the spot. I dragged my poor aching body away from the wall, grabbed the receipt out of Mr Williams' hand, and ran home as fast as I could, meeting my mother in the yard. Before I could say a word, the Ma said, "You'll have to go over to Sheehans and clean up the shit."

"Shag the shit," said I, "are there any more pigs around here?"

The Ma just stared at me, as if I had gone mad.

"How many pigs were there for Donnelly's?" I shouted to her.

When she answered, "Seventeen," I felt my heart going first down into my shoes, and then up into my mouth. I ran around the yard like a madman, looking into every nook and cranny, just in case some of the little hoors were hiding on me. Then I told the Ma what had happened, how only fourteen pigs had gone into Donnelly's. Her consoling words were, "I'd hate to be in your shoes when your father hears about this."

With the Ma's words ringing in my ears, I hopped on my bike and searched all the highways and byways of Cork Street, and its surrounds, but without success. By the time my father came home from work with the horse and cart, I was nearly a hospital case, worrying about what he would do to me. My mother met him in the yard and told him to come straight into the house before he unyoked the horse.

I cowered beside my mother, near the door, out of striking distance from the Da. The Ma told him in solemn tones:

"A terrible thing has happened. It was not the child's fault, and I don't want you to lose your temper over it."

"What has the little bugger done this time?" asked the Da, then adding, "Has he broken another window, or has he been throwing manure bombs again at Miss Gray's Union Jack?"

The Ma pressed on, "The child left here on his own today with seventeen pigs for Donnelly's …" — the Ma paused and looked up to heaven as the Da looked quizzically at her "… and when he got there, there were only fourteen."

For a moment it seemed as if the Da had been struck dumb, then he blurted, sputtered out, "It's only a few yards from the gate to Donnelly's. How in the name of Jasus could himself and Williams lose three pigs?"

"That's the trouble," said the Ma, "he brought the pigs around on his own, he didn't wait for Mr Williams."

The Da said nothing at all for a few moments, and then with a speed of hand which would have put Jesse James to shame, he swept his hard hat off his head and sent it skimming "Odd Job"-like across the room, where it hit me on the back of the head as I darted for the door.

"What kind of a gobshite are you at all?" he shouted, as he headed out the door after me, with the Ma trying to hold on to him. The Ma eventually got him quietened down, and then I was persuaded to go back into the room to give him details of the disappearing pigs. At the end of my tale, the Da said, "So one of them shit all over Sheehan's shop. That'll serve her right." Then he said, "Let's go and look for them pigs."

I ran after the Da as he hurried out through the harness room into the yard, grabbing a couple of ropes on the way. We jumped up into the cart and he drove it right out into the middle of the street, forcing a bus to stop. He went up to the driver and the conductor and explained his problem, asking them to keep a lookout for the pigs. He then drove up to Dolphin's Barn and stopped buses going in both directions along the South Circular Road, again explaining his problem to the drivers and conductors. Next we covered the streets off Cork Street, Donore Avenue, Maryland, shouting to passers-by and people standing in doorways or at bus stops, asking if they had seen any pigs. Back again up to Dolphin's Barn.

At the end of St James's Terrace, coming up to Dolphin's Barn chapel, his eyes suddenly fixed on something just inside the chapel gate. As he stared, I said, "What is it, Da?"

"How often," said he, "do you see pig shit inside the grounds of a chapel?" With that, he climbed down from the cart, and entered the grounds of the chapel. Around the back, quietly lying under a tree, was one of our pigs. Knocking a pig down and tying it up by the legs is a difficult job, but the Da, rope in hand, soon accomplished the task, and fifteen minutes later we were back in the yard to deposit our catch.

On our way back out, we met a neighbour coming back from devotions in Meath Street chapel who said she had seen a pig at the grotto off Pimlico. This was in the opposite direction to where we thought the pigs had gone. We rushed down to

Pimlico, and, sure enough, there was a pig being chased up and down Braithwaite Street by a gang of kids. After beating off the kids, the pig just gave up, exhausted. We got it safely home, but not until the horse had been fed and bedded down did my Da eventually sit down to have his dinner.

But no sooner had he lifted the first forkful than a knock came to the door. It was a bus conductor to say that a pig was wandering around Crumlin Village. We rooted the horse out of its comfortable bed, and off we went to Crumlin Village. At first we couldn't see the pig, but when we got to Captain's Road, there it was, ambling along contentedly on the grass verge. It was almost dark at this stage, and we had to move fast in case the pig would slip away into the darkness, but the Da managed to catch it all right, and then we both hooshed it up into the cart and headed home.

When the Da and I finally were eating our dinners close to midnight, he looked across the table at me and said, "Take those three pigs around to Donnelly's tomorrow."

"I will, Dad," I replied, adding, "I'll get Mr Williams to come around and help me."

"No," said the Da, "you take them around on your own. You started the job, so you'd better finish it, only this time take them around one at a time. St Francis himself couldn't drive three of those around on his own, let alone seventeen of them." Then he added, with a smile, "And you tell me that one of them shit all over Sheehan's shop? I'll say this for you, son, there's never a dull minute when you're around. Here, have another sausage."

A few weeks later, when we were working together in the yard, the Da suddenly said, "That pig we picked up in Crumlin was very dirty, and a bit scrawny looking. That journey up to Crumlin must have taken a lot out of him, he didn't look a bit like one of ours."

THE PHOENIX PARK

The Phoenix Park, covering nearly 1,800 acres, is the largest enclosed park in Europe and the second-largest in the world. For generations of Dubliners it has been a great source of adventure, fun and relaxation. The parts within the seven miles' circumference of the park which interested us most were the Zoological Gardens where we used to go and tease the monkeys, and the Fifteen Acres playing fields where we ran and played football. I never knew why the area was called Fifteen Acres, for there were over two hundred acres of playing fields, including the spot where the Pope said Mass in 1979. This exact spot had been the site of many a bloody death, for it was a popular duelling ground in the eighteenth century.

Sunday was a great day for visiting the Park. Whole families went on picnics to the People's Gardens, where the Mammies and Daddies could smell the flowers and watch the ducks, and the kids could slip across to The Hollow, where they could climb the trees or run up and down the steep paths and hills, while the band played on the bandstand.

The Zoo, situated just at The Hollow, was another favourite family place, with the traders outside selling sweets, drinks and newspaper cones of nuts to feed the monkeys. Many a cone of nuts was bought for the little fellows, but devoured long before we reached their cages. They say that revisiting a place that once gave you happiness and fun is likely to bring disappointment. This is certainly so of my visits to the Zoo in recent years, for the place I remembered going to as a little boy was a happy, cheerful place, where you could ride on the elephants and in the traps pulled by little ponies. The thoughts of going to the Zoo in those days excited me and made me happy. Even though, to paraphrase the words of a great saint, "When I was a child I talked like a child, I thought like a child, I reasoned like a child," I can still make reasoned comparisons between now and then, because even now I am a child at heart. Today there are no elephant rides, and the little trap-pulling ponies have been replaced with a motorised bus and trailer contraption. More profit for the management, but less fun for the children. The place was not all that clean on my last visit there, and unless my imagination is playing tricks, I thought the animals looked unhappy and unwell.

Surely nature parks, with hundreds of acres of land, are far more desirable places of containment for the type of exotic animals you see in small cages in the Zoo. But maybe it's time to scrap zoos and nature parks altogether, to stop taking these magnificent animals out of their natural environment to bring them thousands of miles to confine them in stingy cages and alien climates, where they will lead miserable lives until the end of their days. I have been told that, even if born in captivity, these animals pine for their homes in the wild.

With so many excellent television programmes being made about wild animals in their natural habitat, there is very little left to know about them, and the zoos, apart from letting us see them in the flesh, in all their misery, are not going to teach us as much about the animals as a well-researched and well-presented film. Indeed, going to the zoo, expecting to see magnificent equivalents of animals seen on television, can be a sad anti-climax when one is confronted with dozy, unhappy looking animals, confined in miserable conditions.

The Furry Glen and Knockmaroon areas were popular places for those fancying a long walk from the North Circular Road or Parkgate Street entrances. But with a bicycle, the whole of the park was easily accessible. Our favourite places after the Zoo and the Fifteen Acres playing fields were The Hollow, the army grounds, the dog pond, the area around the Wellington Monument, and the Magazine.

Once the long bright days arrived, we were up the Park several nights a week playing football until it was dark. Sometimes we walked there, but most times we cycled, or were given cross-bar rides. Of course a football was handy if you were going to play the game, but not essential, because with the Fifteen Acres filled with hundred of kids playing football, you only had to ask someone to let you play and you would be given a game. There was an unwritten law among the kids who played football in the Park, whereby no one with a ball ever refused to allow another kid to play.

The footballs used were expensive at the time, and kids had to save as a gang to buy one. Our gang managed to buy a ball or two, which like most others was made of soft leather. This meant that the ball soaked up the rain and mud very quickly on rainy days and became heavier and harder to kick as the game went on. The stitching on the panels of the ball seemed to come apart quite a lot, and as we had leather repair facilities in our harness room, I was often the one elected to stitch the panels. For this I would use hemp which I stretched from a hook attached to the wall, first coating it with wax, and then rubbing in a heavier coating as I rolled it back and forth through a ball of wax over a leather apron covering my knees. Rubber bladders could be repaired with a bicycle tube patch, but sometimes, when it went beyond repair and there was no money for a new one, I would use a pig's bladder from Donnelly's bacon factory.

My pals and I would play soccer in the park for hours on end, evening after evening, in the summers of our youth. And when not playing ball in the Park, we played it up and down Marion Villas. I also played with different groups of pals in Brown Street and The Tenters. We would play until it was so dark we couldn't see the ball, and then we would go home and sleep a sleep that only outdoor sports people can experience and appreciate.

Tommy Moroney, a neighbour, used to define an optimist as a man who took a donkey to Epsom and backed it to win the Derby. I would call him a bit of an ass myself. I became an optimist in my fourteenth year, when playing football one evening in the Park. A gang of kids from a neighbouring pitch asked us to play them in a match. After half an hour or so, they were leading six-nil, which was bad enough, but then they started crowing about it, rubbing it in, and that's about the worst thing a gang from Inchicore can do to a gang from Cork Street. We had been taking the game quite easy up until this point, but in response to the taunts we rearranged our side and set about pulling back the deficit. Much much later, with the score at ten all, and darkness having descended to the point that not only could we not see the ball, but we couldn't even see each other, honour was saved all round. The lesson learned was never to give in, despite the odds, to play until the final whistle is blown.

I also learned another lesson in the Park, to do with either bravery or stupidity, I could never make my mind up which. One summer's evening, myself and a pal were kicking a football around just below the Wellington monument. Close by, there was a group of girls throwing a ball to each other, showing off to us as we were to them in the lovely innocent way that young teenagers did in those days, with the odd glance and the smile. My pal and I had just finished playing and started to take off our football boots, when a large gang of tough-looking kids, coming from the direction of the Army grounds, took the girls' ball, and despite their pleadings, went off with it towards Parkgate Street. To my horror, the lovely big blonde girl, the one I was showing off to, came over and asked me to get her ball back.

I was never particularly good at fighting, and much less so at that moment, when I saw the size of the gang. In a state of mind, which to this day I can only liken to a Japanese Kamikaze pilot, I replaced my football boots and started off after the gang who were still on the grass, heading for the Parkgate Street entrance.

"You'd better keep your boots on," I called to my pal, who later, I discovered, had no intention of following me. I caught up with the gang where the biggest of them had the ball in his hand.

"Hey you, give me that ball," I heard myself call out to him.

His reply struck terror into me, as a passage opened up through the gang leaving us facing each other.

"Are you going to take it off me?"

Then as my heart pounded, my knees weakened, and my mouth started to dry up, I heard myself talking again.

"If you don't give me that ball by the time I count three, I'll spread you all over the bloody park." I knew that the only spreading I could do with any reasonable hope of success would be that of butter on very hot toast. Or put another way, if I were put into a large paper bag, I would have difficulty in fighting my way out of it.

As the big fellow and I stared at each other, and the gang looked on in anticipation of the forthcoming spreading trick, I heard myself threatening him again.

"I'll give you three," and then I began to count out loud, like a referee might have done for the eighteenth-century duellists in the other part of the Park.

"One" — silence, with the ball still held firmly in his hand.

"Two" — how the hell did I ever get myself into this situation for a show-off blonde who doesn't look that good close up anyway?

Just as I was about to say "Three," and take my medicine, the big one suddenly threw the ball on the ground at my feet, saying, "There you are, Kevin Barry, we didn't want the bloomin' ball anyway."

For a moment I was going to tell him to pick it up and hand it to me, so carried away was I with the success of the moment, but thank God my guardian angel overcame the bad thoughts of the Kamikaze spirit, and I simply picked up the ball and walked away, intending to find out more about this Kevin Barry fellow for whom I had been mistaken. Little did I know then, that later in life I would make a good friend of another Kevin Barry.

As I handed the ball back to the smiling blonde girl, I thought how beautiful she looked close up, and how ugly my pal looked still fiddling around with his football boots. Afterwards my pal and I took the long way home by Islandbridge, about two miles out of our way, for fear of running into the gang again.

Years later when I lived in London, I was waiting late one night at a bus stop in Golder's Green for the last bus to Tally-Ho corner. The only other person waiting was a girl. Suddenly a dark gentleman, who at first glance seemed to be about fifteen feet tall, with something of the same dimensions across the shoulders, came up close behind the girl, pushing his body close to hers. This strange land, where everything goes, was not the Phoenix Park, and it was very doubtful if this big fellow from warmer climes would recognise me as Kevin Barry, with all the terror that such recognition might strike into his heart. My guardian angel was telling me to mind my own business, to look the other way, not to get involved. But the little Kamikaze git from Japan was telling me that this could be my wife on the queue, just wanting to get home safely to me.

There were only three of us in the queue, waiting late at night in a dark part of the road. I glanced towards the girl, and saw that the big one, now standing directly behind her, was fast on the way to making medical history as the first Siamese Twin to be joined outside of birth.

While the angel and the Kamikaze git battled it out over what direction my mind should take, I again glanced towards the girl, who at that moment looked at me, appealingly, worried. It was the blonde girl in the Phoenix Park all over again, but this time it was not a summer's evening on my own ground with only the risk of a hiding at stake. Although there was no gang here, my adversary was from a different background, a different culture, whose rules of engagement would almost certainly be different from mine, but, worse still, one who almost certainly would not have heard of the mighty Kevin Barry. Despite all these dangers, the Kamikaze spirit won through, I was about to get involved. I felt my legs taking me the step or two over to the frightened girl, and I heard my voice saying very loudly, very authoritatively, for the big one to hear, "Excuse me, miss, I really don't wish to interfere in your business, but if this big shit is annoying you, just say the word and I'll spread him all over the road and then scrape up the pieces for the police."

These were not my exact words, for there were a few Cork Street adjectives thrown in to gain his attention, to focus his mind, so to speak, to try to convince him that he was now dealing with a tough streetwise adversary who knew what combat was all about. To be honest, what I heard myself say almost frightened me, and I wouldn't have wanted to be in the big fellow's shoes for love or money. But that's the way this fellow Kevin Barry used to get results, and the trick had worked before. My loudly spoken offer of protection was intended to scare off the big one, and it did, for to my great joy and relief, he immediately took off in the direction of Hampstead Heath. For one terrifying moment I thought I was about to shout after him to come back and apologise, the success of the moment having gone to my head. But as in the case of the Phoenix Park incident, beautiful old common sense, of which I have as little as anyone else, prevailed.

The girl, who was visibly shaken by the affair, said, "Thank you, very much."

I must say that she looked good close up, even in the dark. Then she added, "You're Irish, aren't you?"

"How did you know that?" I asked.

And smiling, she replied, "Oh, it was something you said to that gentleman, and the way you said it."

"We Irish are very distinctive, we have a way of sending messages that are clearly understood," I said with a smile.

She replied, also smiling, "That's for sure."

The bus came along, and seated side by side, we chatted until we arrived at Finchley Central, where she left with another "Thank you," and a wave and a smile from the pavement, while I continued my journey home, still a bit shaken, but glad that I had again been mistaken for Kevin Barry, obviously a man of international fame.

THE ANTICS OF ALBERT

Albert was a male cat, mostly dark grey, but with black and white markings. As cats go, he looked a fairly ordinary decent kind of cat, if there is such a thing. But Albert was no ordinary cat, for, as my father said, he was possessed by the devil, and put on earth to persecute him. From my mother's point of view, however, Albert was a little angel, but maybe she felt that way about him because he persecuted my Dad so much. I remember the day Albert arrived, but not the day he departed, or how he departed. But in between times, that cat sure made his presence felt, and exploded the myth that cats have nine lives, for this one had many more than that.

Our yard was a place for all kinds of animals, including strays. Stray cats and dogs, and even a sheep, a cow, and the odd hen from some neighbour's yard, dropped in at one time or other, and stayed long or short periods before moving on. Albert was one of those animals which strayed in, as was Rambler, a Wicklow Collie sheep dog which rambled in and out over a period of months, before choosing the good life of our farmyard (hence his name). We also had two other dogs, Flora, a small Corgi type, named after Flora McDonald, one of my mother's heroines, and Rover, a massive dog. Until Rover came into my life, I had never seen an animal of such size, power and majesty of movement. Over the years, I tried to identify the breed, but without success, eventually putting it down to one of those quirks of nature. Then, recently, I met a man out walking his dog on Killiney Head. It was Rover, down to the last detail. He told me that it was a Karabash, a Turkish sheepdog, and that the breed had been introduced to Ireland in 1970. Rover, if he was a Turk, must have been some kind of a refugee, for he figured in my life long before that date. When my father got him from a travelling man, he was only a few months old, and looked like a big fluffy ball of wool. But he grew, and grew, and then grew some more. But more about Rover elsewhere.

Early one Sunday morning, my mother opened the front door on her way out to Mass with me, and tripped over kitten Albert asleep on the doorstep. She fell flat on her face, sprawled half on the footpath and half on the road, with legs and arms spread-eagled, and her upper dentures somehow ending up on her handbag beside her head. A new kind of stray had arrived.

My Da rushed out and helped Ma back into our kitchen where he put her sitting shaking by the fire as he cleaned up her bloodied face. I carried in her handbag, but refused to go near the teeth, which even in the glass by her bedside at night, or tucked away at times on the top of the wardrobe, looked positively dangerous.

When my mother recovered, Albert, who had followed us in, was by her legs looking up at her and meowing. If it was me who had taken such a fall, I would have kicked the head off the little hoor. But not my gentle mother.

"Come here, luv, you didn't mean to trip me, did you?" she gummed to him. She said he looked hungry, and so she fed him, and fed him some more, and even more again. Then she bedded him down in a cardboard box in a corner of the kitchen. The same day she named him Albert, after the son of a relation on Dad's side of the family. If Albert's IQ could have been measured, it would probably have fluctuated between minus ten and plus one hundred and fifty. Put another way, if Albert was human, blonde and female, she would be the dumb blonde in all those films who got kicked around, but still ended up with the fur coat and Rolls.

Albert was lazy without being fat, and into meat and milk in a big way. He ate mostly from the pig swill trough. But if he had to go out and catch his meat, as in our furry little rodent friends running all over the place, I think he would have become a vegetarian. Albert didn't mix too much with the other wild cats of the rooftop, preferring mostly to sit and plan the next load of chiffon he was going to land my Dad in. He did not appear to be a sexually active cat, but there was the odd morning when he had a suspicious looking smile on his face and a bit of a jaunt in his step. Albert, like all other cats, could climb and jump on to ledges and window sills, but, oddly enough, couldn't jump down easily from any height above four or five feet without first kicking up the mother and father of a racket meowing. Then he kind of tumbled off the ledge, landing in a painful looking heap, which was strange for a cat. But Albert knew that if he kicked up a meowing racket for long enough, more often than not someone would lift him down.

We had a tall young sycamore tree growing up against the wall in the yard, and one day Albert climbed up to the top but was unable to get down. Eventually the meows started. They went on and on, bringing in a concerned Mrs Horan, one of my Dad's tenants who loved cats. Eventually, at the pleading of my mother, the Da went to get a ladder. In the meantime I started shaking the tree back and forth at an ever-increasing speed, until eventually Albert was whiplashed out of the tree, straight into the middle of the five-foot-wide drinking trough. That was the day I discovered that cats don't swim very well, and horses won't drink water from a trough where tomcats have splashed. Albert was dried and brought indoors to recover, while the Da left back the ladder and gave me a pat on the head and said, "Well done, son."

The next adventure was when Albert, making his way along a series of gradually ascending roofs, finally got on to the apex of our house roof. Once up there, he started meowing to get down. We were all inside the house and couldn't hear him, when Miss Gray, a neighbour whose back yard adjoined ours, and Mrs Horan came in to tell us about him.

"Let him jump," said the Da.

"If that poor little darling jumps and is killed, I'll never forgive you," said the Ma.

"Seems fair enough to me," replied the Da. Then an unmerciful row broke out about Albert. "If you think I'm going to climb up on to that roof and take a chance on killing myself for that shaggin' cat, you have another think comin'," said the Da. While the row was going on, Miss Gray, thinking that Albert was in danger and the Da wasn't going to rescue him, went down to O'Grady's pub on the corner of Ormond Street, to ring the fire brigade to rescue Albert.

Apparently, what she said to the fireman was, "Come quick, Albert's on the roof and we can't get him off. If you don't come quick, he'll fall, or he might even jump. He has jumped off heights before and hurt himself."

To rescue Albert, the Da first had to climb up onto the roof of our kitchen by ladder, drag up the roof ladder, and then the climbing ladder that got him up onto the roof of the kitchen. Next he set the climbing ladder up against the main house wall and climbed up, dragging the roof ladder behind him. When he got to the top of the wall, he manoeuvred the roof ladder, while still standing on the climbing ladder, and clipped the special end of the roof ladder over the roof tiles on the apex, to keep it from slipping off the roof. After a lot of huffing and puffing, he very carefully got onto the roof ladder, and started to climb up the roof towards the meowing Albert. But just as he reached out to Albert, the cat turned and walked away, out of his reach. Suddenly, to our horror, the clip broke off the end of the ladder, and it, with the Da attached to it, came sliding down off the house roof and crashed onto the kitchen roof. All of us were sure the Da was killed, for he lay motionless on the roof, still clutching the ladder. As the climbing ladder was still on the kitchen roof, propped up against the house wall, we had no means of getting up to reach the Da. Mickey Joe and Kevin from next door were quickly summoned, and they climbed out through the landing window, onto the roof, and gently passed the ladder down to ground level, the Da still lying on it, gripping the rungs of it tightly with white-knuckled, blood-stained hands. As the Da was helped to his feet, obviously suffering shock and pain, the neighbours gathered around, delighted that he was all right. Then my Dad was clearly heard by all within an ass's roar of him:

"That fucking cat will be in the dung heap today, with his head stuck up his arse."

This explosion of bad language, coming from a normally well-controlled quietly spoken person, who hardly ever used a swear word, was just too much for the women neighbours.

Mrs Horan immediately launched into "Hail Mary, full of grace …" and dropped to her knees blessing herself. Unfortunately she forgot that one of her legs was considerably shorter than the other, and in the heat of the moment, not allowing mentally for the physical adjustment necessary, she toppled over sideways, changing her prayer as she fell to,

"Holy Mother of God."

With that, Mickey Joe, Kevin and Harry, who had just arrived, burst out laughing, and although the Ma and me at first tried not to join in, because of the possible repercussions from the Da later on, we nevertheless succumbed, and just let go, laughing like everyone else, while the Da just stood there repeating his threat about Albert. Mrs Horan and Miss Gray were not amused, however, and they hobbled off together, one holding up the other, with Mrs Horan still praying aloud.

The rest of us, still trying to hide our laughter but without too much success, got the Da into the kitchen, gave him a cup of tea, and put him into an armchair and covered him up with a blanket. Then we washed, iodined, and bandaged his poor hands, and did the best we could with his face, which was scraped. When quiet had descended on the house, except for the distant meowing of the cat still on the roof, the Ma suggested that Da should go to the hospital for a check-up. When he replied, "What, and not be around when that stupid cat falls down off the roof?" the laughter burst out all over again, with the most laughable and laughing bunch of people I have ever known.

It was only the sound of the fire engine arriving that brought things back to normal. We rushed to the front door, the Da hobbling after us, to see the activity.

As the firemen jumped off the running board of the engine, Dad said, "There must be a fire in the furniture factory."

By now, passers-by were gathering and a fireman was heard to shout, "Where's the man on the roof?" Mickey Joe, putting two and two together quicker than anyone, which was quite a trick for him, told the fireman that it was a cat on the roof.

"The woman who telephoned said Albert," replied the fireman.

Just then Miss Gray arrived to say, "It is Albert, Albert is the cat's name."

The fireman again, "You mean to say you brought us out to rescue a cat? And what bloody eejit would give a name like that to a cat?"

We all looked at the Ma, who stepped forward, and very apologetically said, "Well, he does resemble a relation's son. He's living in Chicago, where he drives a bus, and I heard he's doing very well, making a lot of money,"

The fireman: "It's a pity he isn't living here and could drive a fire engine, then he could have gone on this bloody wild-goose chase instead of us."

By this time the fire engine ladder had been raised, and, to the cheers of the onlookers, a fireman climbed up to fetch Albert. But just as he got within reach, Albert did his walking trick again, moving out of range. The fire engine was repositioned, and the ladder swung to the far end of the roof. Again the cheers, and again Albert walked. After another ten minutes or so of this fireman climbing and Albert walking lark, the head fireman said:

"That's it, lads, let the stupid little bastard stay up there." Then he added, "The easiest way to get him down is to shoot him down."

"Have you got a gun?" asked the Dad, becoming excited at the prospect of Albert being lead poisoned — through the head.

"I wish I had," replied the fireman, before jumping onto the running board with the rest of his crew and driving away.

Some people called up "Pussy, pussy," and Albert meowed back down to them. After a time he came to the very edge of the roof and bent down, hunching himself, as if getting ready to jump.

"Go back, go back," chorused some onlookers.

"Here pussy, here pussy," called the Da, wanting him to jump. Then Mrs Moroney came out with a big rug, calling people to hold it at the edges so that Albert could jump into it. It seemed stupid to imagine that the sight of an open blanket would entice a cat to jump into it, but coincidentally or not, Albert did jump, to a big cheer from the onlookers. But as quickly as the cheer rose, it dropped to a stony silence, for Albert had missed the blanket and landed "splosh" on the pavement.

I was the first over to pick up the lifeless bundle of fur, and I carried it into the yard, closely followed by a broadly smiling Da.

"Are you satisfied now?" shouted my mother to the Da, as the tears rolled down her cheeks. Although in pain himself, he put his arm around her shoulders to comfort her, as she continued to pump out tears of sorrow, which only a short time earlier had been tears of joy, albeit at the Da's expense.

"What will I do with Albert?" I said to the Da, as he stood with his arm around the Ma with the neighbours looking on.

"Sling him in the dung heap," he replied.

With that the Ma pulled away from him, shouting, "You should be in the dung heap, the whole bloody city is nothing more than a dung heap, and I curse the day I ever came south to it."

This brought a round of applause from the onlookers, and someone called out to the Da, "Throw yourself in the dung heap."

Things were not going the Da's way. Then he said quietly, "Bury him under the sycamore tree."

As we moved up the yard, the crowd moved off. I laid the cat down on the ground, and went over to Sheehan's shop to get an empty sweet box to bury him in. Little devil and all as he was, he deserved that. The box was a bit small, and so Albert had to be stuffed in, with the lid on slightly skewways. The Da took the shovel and dug the hole under the sycamore tree, as the Ma looked tearfully on. After digging down about a foot, he reached for Albert's coffin, only for my mother to stop him by saying, "Deeper." He dug down another foot or so, and again reached for the coffin.

"Deeper still," came another command from the Ma. Again he dug, only this time, to get added depth, the hole had to be big enough for him to stand in, in order to get leverage on the shovel handle. By now the hole was nearly three feet wide, and at least two feet deep.

"Is that deep enough for you?" asked the Da, as he reached again for the coffin.

"A little deeper still," said the Ma. With that the Da exploded.

"Jasus, woman, it's not Robert Emmet we're putting down here, only a stupid, mangy old cat, that has caused nothing but trouble since he bummed his way into the house. I've fallen off the bloody roof trying to rescue him, the fire brigade was called out, Mrs Horan has threatened to bring down the parish priest on me because I lost my temper, and Miss Gray says she will never talk to me again. Now let's bung the little bastard down the hole and go and have a cup of tea."

My father was in pain and terribly frustrated, while my mother was heart-broken over the loss of Albert. As Da apologised and put his arm around the Ma once again, I took over the shovel and dug down another bit. Then I reached for the coffin and held it out at arm's length in front of the Ma, for a final farewell.

"Get on with it," snapped the Da.

At that point the cat's tail slipped out of the box, pushing off the lid. I tucked the tail in again, and put back on the lid. Again the tail came out, only this time, more popping out than slipping out. Again I tucked the tail back into the box, and bent down to pick up the lid that had fallen onto the ground. When I bent down, Albert fell out of the box onto the ground, and started to wiggle, trying to stand up.

"He's alive, he's alive!" shouted my mother, as she bent and picked up the cat, and hurried indoors cuddling him in her arms and kissing him.

"I don't understand this," said the Da. "I fall off the roof, nearly kill myself, and everybody breaks their sides laughing, not giving a damn about me. In my condition, I then have to dig up the bloody yard, to give the bloody cat a State funeral, and just as we're getting to the best part, when we're about to get rid of the little hoor forever, he does his Lazarus trick, and is carried off hugged and kissed." Then he said, "It's all your fault, you know."

"Why is it my fault?" I asked incredulously.

"If you hadn't been doing the theatrical bit, holding him out to your mother for her final blessing, but had just bunged him down the hole with a few shovels of clay to keep him warm, he'd have been forgotten about by now, and we'd be inside having a mug of tea and few Goldgrain biscuits," he said.

———

And then there was the episode of Albert and the Rat.

One evening as the Ma, Da and I sat by the range fire in the living room, the Ma reading the *Evening Mail* to Da, while I read the *Hotspur* and listened to Donald Peers singing *In a Shady Nook, By a Babbling Brook* on the radio, Albert came in through the open door and stood with something in his mouth.

"Look," said the Ma, "Albert's got something nice to play

with."

"What is it?" asked the Da.

"It's a dead rat," said I.

With that Albert dropped the rat in the middle of the floor, turned on his heels and walked out. The rat rolled over and scuttled under the Da's chair.

Like a flash, and with a kind of scream, the Ma and I were up on the couch, while my Da, equally quick on to his feet, shouted, "Quick close the door. Don't let him out."

"Close it yourself!" shouted back the Ma, as the two of us clung to each other. My Da slammed the door, and, clever man that he was, put on his leggings to make sure that the rat didn't find its way up his trouser leg. A cornered rat had once gone up the trouser leg of a relation, biting, and caused him much pain and a long illness. Then the chase started. The rat ran under the table, then under the couch, behind the side wall cabinet, and then behind the dresser. My Dad followed it around the room, flushing it out of every hiding place with the poker poised ready to strike. But no sooner did the rat vacate one place than it moved to another place, doing the rounds all over again.

Whereas the table, side wall cabinet, and even the couch with my mother and me on it, could be moved without too much trouble, the two-piece dresser, filled with plates, jugs, and hanging mugs, was very heavy and hard to move. But as the rat was under or behind the dresser, it had to be moved. Unfortunately, in moving it, my Dad allowed the top half of the fully-laden dresser to come crashing down on the floor, breaking every plate, mug and jug that it contained.

By this time the Da had lost the head altogether, cursing and swearing as he trampled over the roomful of broken crockery, finally sliding on it and falling heavily on his back, while my mother broke out in hysterical laughter. As the Da struggled to get up, at the same time lashing out with the poker as the rat ran past him, his feelings were clearly expressed:

"It's not this poor bastard of a rat I should be chasing, but that stupid bollix of a cat."

I can honestly say, with hand on heart, that until Albert arrived in the house, I never heard my Da curse. I didn't think he knew how to, but he did, most gloriously.

A little door at the bottom of the range had been left open for ventilation, to kindle up the fire and allow extra heat to come out through the top. It was in through this door that the stupid little rat ran, with the Da quickly closing the door behind him. This brought the chase to an end. The rat burned to death, its sickening smell filling the room.

"That bloody cat's for it now!" shouted the Da as, poker in hand, he waded through the broken crockery and out through the door looking for Albert, with the Ma after him pleading for the cat's life. But Albert, operating at the highest level of his IQ, was nowhere to be found that night, or for a few following nights, until the Da had quietened down.

In many ways Albert and I had a lot in common, both of us causing trouble and then having to stay out of the Da's way for a time.

———

Let me tell you about Albert in the Cellar.

My father was not the only one to suffer physical pain because of Albert, for I too had an experience, which again involved the fire brigade. Rambler, one of our dogs, was a gentle animal who loved people, and loved to be patted on the head. One of Rambler's favourite spots to sit and be patted was at the bus stop opposite our house. He soon became a favourite with the regulars who queued there, particularly the nurses from Cork Street Hospital. If there was no one at the bus stop, Rambler would just sit at our gate, or on the step of the cake shop. But as soon as someone arrived at the bus stop, he was over to sit beside them, and then to stand up and wag his tail at them as they got on the bus. He was just short of shaking hands with everyone he met, a regular Alfie Byrne of the dog world. Alfie was the Lord Mayor of Dublin who shook hands with nearly every one he met.

Albert and Rambler developed a special relationship, and there was many a hot summer's day when Albert could be seen cuddled up inside Rambler's paws as they both lay fast asleep in the yard.

"Oh, isn't that lovely?" my mother would say to my father, as they both looked at the sleeping beauties. "Get the Brownie and take a picture of them," said the Ma.

"I'll do that," said the Da reaching for the shovel, "but first let me give Albert a little tap with this on the head, so that he doesn't move when I'm taking the snap."

With that my mother turned on him and followed him around the yard giving him a good tongue-lashing while he just laughed.

Albert soon took up the habit of sitting at the bus stop with Rambler. But whereas Rambler would never get on to the platform of the bus, unless going somewhere special with me, Albert, on the other hand, developed the habit of jumping on and off the platform of the bus. In the beginning the bus would go only a short distance before pulling up to eject Albert. But then, as the bus conductors got to know him a little better, he was allowed to ride to the next bus stop at the corner of Ardee Street, where he was put off to walk back. This happened time and time again, and it was no surprise to see Rambler and Albert sitting side by side at the bus stop with a big queue of people, and then when the bus had pulled away, only Rambler would be there, Albert having gone for a ride. A few minutes later, Albert would arrive back beside Rambler, having returned from the Ardee Street stop.

But one day Albert was ejected from the platform of the bus in what, judging from an eye witness's description, was a very positive and unceremonious manner. It was a Saturday. The Ma, Da and me were having our dinner at about one o'clock, when Mrs Moroney, a neighbour and friend, burst into the kitchen shouting, "Quick, quick, Albert's in Breen's cellar, and I think he's dead."

The Ma and I were up and out like a shot, across the road to look into Breen's cellar. True enough, there was the poor Albert, laid out on the floor of the cellar with no apparent sign of life in him.

"Oh no," said the Ma, with her two hands up to her mouth, "poor Albert."

She looked around for my father and discovered he wasn't there. Then she strode back to the house with me after her.

"What kind of a man are you, who'll sit there eating your dinner, with poor old Albert lying dead or unconscious in Breen's yard?" she asked him.

"A hungry man, who doesn't give a damn about that stupid cat," he replied.

After an argument, the three of us returned to Breen's cellar, where by now a few women and kids had gathered around to look down at Albert's body.

"What happened?" my Da asked Mrs Moroney.

"Well," she replied, "I was walking down the street and saw some people getting on to the bus. Suddenly Albert shot off the platform of the bus, at a height of about four feet. He cleared the railings, smashed off the wall of Breen's house, and dropped into the cellar."

"Some terrible bus conductor must have thrown him over the railings, he should be sacked," said the Ma.

"Some very sensible bus conductor has kicked the little hoor off the platform of the bus and he should be given a medal," said the Da.

Breen's cellar had railings around it about four feet high. The floor of the cellar was about six feet below pavement level, which meant that when Albert came over the top of the railings, he then fell at least ten feet. As the railings had no gate, the safest way to enter the cellar was to squeeze through the two bars of the railings which had been set a little bit wider apart than the others. To climb over the top of the railings and drop down, or just to climb over the top and climb down, was a bit risky, for a fall on the spiked railings could kill you.

"Get the cat," said the Da to me, and dutifully I put my head through the bars to climb down into the cellar, something I had done many times over the years, but not in the last few months. I found it very hard to get my head through on this occasion, particularly the ears, but after a lot of pushing and twisting, I managed it. Next I got my right leg and right arm through the rails, but no way could I get any more of me through, no matter how hard I tried. The fact was I had just got too big to fit through the railings.

When this was eventually realised, the Da said, "Climb over the top, son."

But alas, my head was firmly stuck in the railings. I had had great difficulty in getting the head through in the first place, but try as I might, there was no way I could get it out. By now it was me and not Albert who was the centre of attraction, for a crowd had gathered round to look at the stupid kid who had got his head caught in the railings.

After about fifteen minutes of trying to get my head out, during which time sympathetic voices told me not to panic, while some little fecker of a kid pinched the bum off me, it was decided that the fire brigade should be called, and someone went down to O'Grady's pub to make the call. In due course the fire engine arrived, complete with helmeted firemen standing on the running board, and captained by the same fireman who had been in charge of the Albert rescue attempt.

"What, no cat on the roof today?" said he as he surveyed the situation, adding, "At least you have a good reason for getting us out this time, and not because of a stupid cat with a stupid name. Hold on there, son, we'll have you out in no time."

In no time was right, for quickly a fireman placed some kind of gadget between the two bars which held my head, turned another bar at the side a few times, and, like magic, the bars holding my head widened right out enough to let me get free. But as I came free, and the onlookers gave a big cheer, instead of pulling my head back from the railings, I pushed it and the rest of my body through and jumped down onto the cellar floor to get Albert's body.

In the excitement of the fire engine arriving, no one had noticed that Albert had done his Lazarus trick again, and was sitting up meowing as I reached him. I grabbed the cat and climbed back up to the street, where I handed it to my Dad. He in turn glared at it, and passed it over to the Ma who immediately took it indoors.

"Just imagine," said the Da, "her own darling son gets his head stuck in the railings trying to retrieve that stupid cat, and when he gets out, not even a hug for him. Come here, son, I'll give you a hug. And I tell you, that stupid cat is nearer the dung heap today than it has ever been."

The fire chief, noticing the hand-over of the cat, and overhearing the Da's comments said, "Don't tell me that that bloomin' cat is somehow involved in getting us out here again today?"

Then my dad told him the whole story, after which, in a very annoyed voice he said, "The next time I'm called out here, I want to see at least smoke, if not flames to go with it. What I don't want to see are stupid cats on roofs, in cellars, riding on buses, sitting in window boxes, or crossing the road, unless a

THE HOUSE OF RECOVERY AND FEVER HOSPITAL

We lived about 100 yards from the House of Recovery and Fever Hospital, and the Weir Nurses' Home which faced it. The hospital, the nurses' home, and the adjoining graveyard, together with the nurses and the hospital porters, were a great source of fun and adventure as we grew up. But few of us knew the history of this marvellous hospital, which served the poor of Dublin for over one hundred and fifty years.

Dublin, in the time of the Georges, with public buildings such as the Houses of Parliament and the Custom House, and its elegant streets and squares, was a city of fine architecture, and one of the most beautiful in Europe. For the well-off minority, it was a city of culture, gaiety, drama, art and music. But underneath this glittering facade, and a stone's throw from the finest buildings, there existed some of the worst slums ever seen. Sanitation was appalling. The reservoirs which supplied the city often contained dead animals. There were no sewers. Cattle were slaughtered in the streets. Cesspools regularly burst, adding to the prevailing stench.

Such was the accumulation of offal in the streets, that people sometimes had to wade knee deep through the filth to get across. Dublin had nineteen cemeteries in which the bodies were buried in shallow graves. Add to this the appalling housing conditions of the slums and the inevitable result was widespread disease — enteric and cholera, typhus, smallpox, to name but a few.

It was against this background that a group of Quakers, joined by the Huguenot family of La Touche, and later by members of the Guinness family in 1794, established "The Meath Street Institution" which aimed to provide medical aid to the sick poor, to assist them and their families with the necessities of life, and to prevent the spread of communicable diseases in Dublin. Methods of prevention included the fumigation and lime-washing of the rooms of infected people and the destruction of infected bedding and clothing. Such was the origin of preventative medicine in Ireland.

In 1801 the same group decided to build a House of Recovery and Fever Hospital, and having collected £1,375/12/6d, they bought a plot of approximately three acres, in almost the highest ground near Dublin, "the orchard of the widow Donnelly" in Cork Street. The hospital welcomed its first patients in May 1804.

Over the next century and a half, the hospital grew and developed in response to the changing needs of the growing city. Troops returning from the Napoleonic wars brought exotic fevers and pestilence to add to the noxious brew of cholera, typhus and later diphtheria, epidemics of which raged throughout the nineteenth century. The demand for the hospital's services was sometimes so great that emergency beds would be set up in tents in the grounds. In 1903, the Weir House, considered one of the finest nurses' homes in Dublin, was opened directly across the street from the hospital. Mr James Weir, a retired business man who had been nursed through a serious illness by a Cork Street nurse, provided in his will for the erection of the home. By 1953, however, the hospital had outlived its original purpose, and it was closed down.

In its time, the House of Recovery and Fever Hospital, founded by those humanitarian business men of Dublin, contributed greatly to the alleviation of suffering and pain. But for us snotty-nosed kids from Cork Street, fortunate not to have known the horrific underside of the much-celebrated Georgian period, the hospital and the nurses' home were nothing more than a source of adventure and fun.

The hospital played a big part in the lives of the surrounding community. The constant stream of staff and visitors provided extra custom for the local shops. Almost all of the nurses were beautiful young country girls with lovely accents, many of which we Dubliners were hearing for the first time as they walked back and forth between the hospital and the nurses' home, in their crinkly starched uniforms, laughing and joking among themselves. They were always cheerful, and courteous to anybody they met. I suppose one of the reasons why Irish nurses are so well liked around the world is that their good humour and friendliness alone are enough to cure you, apart altogether from their skills. I think it fair to say that Dubliners at that time tended to be somewhat unfriendly towards country people for all kinds of silly reasons.

Sometimes, on the day of the All-Ireland final, we would go down to O'Connell Street early in the morning to see the country people walking around like lost souls, with their trouser bottoms rolled up over their boots, wearing their team colours and talking in a dialect that only they understood. I dare say the country people thought us funny-speaking Dubliners were as strange as we thought them. One thing for which I am grateful to those young nurses from Cork Street , is that, through their courtesy and sense of fun, they gave me a love for all people from every part of our lovely land. The important thing is that, first and foremost, we are all Irish people, not "county" people, and when the real everyday problems of this land need to be solved, it is only we who can solve them, not the strangers.

The nurses' home had about two acres of grounds, which ran parallel between Cork Street on one side and Marion Villas on another. On the third side, it backed up against the wall of the Scribbans Kemp cake factory. The fourth side ran the full length of the rear of the nurses' home building. The entire grounds were surrounded by a high limestone wall with just one gate on the Cork Street side. In the grounds, tight up against the Scribbans Kemp wall, was a beautiful old wooden summer house. Beside this was a tree which could be climbed from the summer house, and gave access to the Scribbans Kemp roof. The ground itself was an old Quaker's graveyard, with no more than ten graves. The gravestones were dotted here and there about the ground. One name that I remember was "Sophia Webb," who was only a child when she died. The headstone for "Old Soapy," as she became known, was free-standing, as were one or two others. On occasions, the gang moved some of the headstones into position to make goalposts for a quick game of football. And believe me, it had to be a quick game, for the porters were always on the job, ready to chase us, although they never caught up with us.

The hospital stood on about three acres, also surrounded by a limestone wall. The wall was about eight feet high, but just opposite Marion Villas, it tiered down to about six feet. It was at that point we could climb up easily either to drop into the grounds when up to mischief on dark evenings, or just to sit on the wall on a summer's evening and watch the nurses playing tennis. Nobody went too near the mortuary. This building inspired a heady mixture of excitement, fear, and horror. But it was in and around the grounds of the nurses' home that we were able to enjoy the fun of talking and joking with the nurses, and the excitement of being chased by the porters.

Marion Villas, off Cork Street, is a cul-de-sac, with the wall of the graveyard on one side, a few houses on the other, and a factory gate at the top. Some of my pals and their parents, all of them lovely people, lived in the Villas. Night after night, year after year, as we grew up, playing football, racing, cricket, and umpteen other games throughout the long summers of our youth, the adults must have been fed up with all the noise, but never once did any of them complain.

These days nearly every kid has a football of one kind or another, and the supermarkets and shops are full of them. When we were growing up, possessing any kind of a ball was wonderful. Very often we made balls from newspapers tied with string, but of course they didn't last long. The best kind of ball to have was a tennis ball, for with that you could play soccer, or cricket with a hurley. Very often the ball would go over the wall into the graveyard, and to make it easier to get it back, we made grips at three points in the wall, at both ends and the middle.

We liked to climb over the graveyard wall, and slip across to the nurses' home for a chat. We could only do this when it was dark, for fear of being caught by the porters. Even then, we had to be careful and put someone on watch. We would climb up onto the high wide window sills which looked out onto the graveyard, and all the lovely young country girls would open the window and gather around on the inside, talking and joking with us, and giving us sweets.

"Don't they talk funny," said one nurse with a strong Cork accent.

They would tell us their names, and we would tell them

ours. Sometimes away from the nurses, we would shyly say the names of our favourite ones, and with the mention of each name, a big cheer would go up. There were always some nurses that a few of us liked and wanted to claim as our own, but rather than fight about them, for we had more important things on our minds like playing football, we agreed to share them.

We got to know the nurses very well, and they would call to us as they crossed the street:

"Hello Paddy, hello Brian, hello Leo, will you be at the window tonight?"

But one thing we could never forget about the grounds of the nurses' home, despite the fun we had with the nurses, was that it was a graveyard. When the nurses closed the window, and it was time to go home, there would be an unmerciful dash for the wall, with the faster ones yelling over their shoulders:

"Look behind you, Old Soapy is catching you!"

Once at the wall it was murder, with body over body, fighting and pulling each other back with shrieks of fear, as we scrambled to get over and away from "Soapy."

The porters nearly caught us twice. On one occasion, as we chatted at the window on a dark winter's night, a group of porters rushed out the hospital gate, some going up Marion Villas to cut off our retreat over the wall, while others watched the Cork Street gate, and more headed into the nurses' home itself. But thanks to an alert lookout, who gave the secret yodel, we managed to escape via the roof of the summer house, over the Scribbans Kemp roof, and out on to Marrowbone Lane. Then we strolled down Cork Street to Marion Villas where the porters were running up and down in a state of excitement.

When we asked one of them what all the fuss was about, he replied, "At last we've managed to surround some of your gang in the grounds, and tonight we'll have the little brats down in Newmarket." (Newmarket was the local Garda station.)

The second lucky escape was when myself and a pal were perched on the window, and one of the nurses shouted, "Quick, jump in, the porters are on their way."

Right enough, the clever ginks had climbed the Marion Villas wall, and seeing our silhouettes at the nurses' window, had jumped into the grounds and were racing across to catch us. Quick as a flash, the two of us jumped into the nurses' sitting room, and they quickly closed and shuttered the windows behind us. One of them rushed us out towards the hall. But just as we were making our way towards the main entrance and freedom, through a glass door we saw two porters approaching. A quick-thinking nurse pushed us in to a nearby laundry cupboard and closed the door behind us. Shivering with fright, we heard the porters hurrying by, heading straight for the sitting room, from which by now floated out the sound of the piano playing *Always*, and the nurses singing along. Whenever I hear that song, it always brings back the memory of that adventure, as well as another special time in my life.

The porters could not get into the sitting room, because the nurses had locked the door. Thinking that they had us trapped in the room, the porters banged on the door and shouted at the nurses to let them in. But the harder they shouted and banged, the louder became the strains of *Always*. By now, all the nurses had joined in. Eventually the banging and shouting forced the nurses to abandon the Irving Berlin classic, and call out to the porters that they could not find the key to open the door. One of the porters then bellowed:

"You're protecting those little brats, but we'll get them."

Then he told the other porter to wait there while he went for a spare key. As we trembled no more than twenty feet away, we bravely peered through the door of our cupboard to see what was happening. Just as the porter went out the door onto Cork Street on his way to the hospital for the spare key, a group of nurses came in through the door, leaving it open. We nipped out and raced home as fast as we could.

A few days later, the nurses told us that the porters were flummoxed when they got into the sitting room and found none of us there. Those lovely girls had saved our bacon, and knocked a bit of fun out of the incident. The funny thing was that although for years we went back and forth over the wall, yapping to the nurses, and getting into all kinds of other divilment, the porters never got close enough to be able to identify us. Many a day, after a big chase the night before, I would stand at our gate as one of the porters who had almost caught me the night before passed by. He always greeted me: "Hello there, son."

We did not venture too often into the hospital grounds itself, for the only attraction there was the mortuary, tucked away at the Brown Street end. And with those grounds crawling with porters, there were definitely friendlier places to be. But my pals and I did brave the territory a few times, trying to see some bodies inside the dead-house, as we preferred to call it. But each time we were disturbed by the living — the porters — and had to abort our mission, and withdraw over the wall into Brown Street. One of my pals told me that he did manage to get into the dead-house one night, and according to him, there was a full house and it wasn't a poker hand …

Many's the night, alone in my bed in the darkened room, after a graveyard or a dead-house adventure, I got the shivers thinking of where I had been, and what could have happened to me. But come the daylight, all my fears of the previous night were forgotten, and with my pals I was planning another trip over the nurses' wall, to see Kathleen, Nancy, Maureen, Áine, Gráinne, Teresa, Pauline, Margaret, Maria, and all the other lovely country girls, who gave us sweets, laughed and joked with us, protected us when the chips were down, and even sang *Always* for us.

ROVER AND THE CORNERBOYS

It is mostly young people who are categorised as trouble makers, sometimes rightly so, but sometimes just because of the way they dress, the kind of music they listen to, or how strongly they express their views — views which our highly educated young people of today can express very clearly and forcefully, particularly to our decision-makers. But sadly the decision-makers have stopped listening, to the young and the not-so-young, and the frustration of this has made them angry. The young undesirables of today are called gurriers, skinheads, layabouts, and other names. When I was growing up, they were called cornerboys, bowsies and lousers, and the women were called brazen hussies.

Cornerboys were so called originally because they hung around at corners, an innocent enough pastime, and understandable considering the social scene which prevailed at the time. In the 'Forties particularly and on into the 'Fifties, there always seemed to be masses of people on the streets, coming and going all the time, on trips of necessity during the day, and on pleasure in the evenings, especially the summer evenings. Lots of people cycled, for the bike was then the main means of transport, but, for socialising, a nice long walk with a group of friends on a summer's evening took a lot of beating.

Before television became the mental disease it is today, imprisoning people in their homes and softening their brains, people visited each other more often, to chat, play cards, tell stories, have a sing-song. Groups of friends going walking would sometimes link arms and sing songs as they walked along. From Cork Street, the favourite walks were through Maryland, out on to the banks of the canal, past Rialto Bridge, and on from there for as far as you cared to walk. Or to Dolphin's Park, just over Dolphin's Barn Bridge, to sit on the grass and watch the football matches. Other favourite places to walk and linger awhile, were the Wellington Monument, The Hollow, and the Furry Glen in the Phoenix Park. Those not so inclined to exercise would gather at their favourite meeting places, usually a corner, and there they would chat to each other and to passing friends. Sometimes the boys would try to get on chatting terms with passing groups of girls, out for an evening's walk.

Nearly every street in the area had its favourite street corners, and its fair share of cornerboys to adorn them. Not all cornerboys were welcome at a particular street corner, however, sometimes because of the noise they made until late at night, and sometimes because of how they would mess up the corner in different ways.

Parking their bums on parlour window sills also caused problems, for not only did it exclude the daylight from the room, it also blocked the view from behind the curtains. For the latter reason, particularly, many owners of corner houses had small spiked railings specially made up and bolted to the sills. As well as the cornerboys watching life's passing parade of humanity, the Mammies and the Grannies were also watching, but from behind their lace curtains, sometimes taking their meals there for fear of missing out on some street activity. But whereas the majority of cornerboys were decent fun-loving fellows, some areas threw up the troublesome kind, whose remarks to some passers-by would quite often cause a fight, branding the whole cornerboy fraternity as trouble-makers, bowsies, or simply cornerboys in the worst sense.

One summer evening, when I was about eleven or twelve, I was coming home on my own from the direction of the South Circular Road, taking a short cut to Weavers' Square through a passageway which ran between Susan Terrace and O'Curry Road, past Hannigans, the little greengrocer store. Just as I was passing the junction of Weavers' Square and Brown Street, one of a group of cornerboys aged about eighteen or so gave me a belt on the back of the head for no reason at all. Another one tripped me up, sending me crashing to the ground. When I got up, another one gave me a belt in the face, sending blood spurting from my nose, and at the same time another one kicked me. Then I had my arm twisted high up behind my back, and one of them marched me up and down in excruciating pain while the others just laughed. Eventually I was thrown into the centre of the street, and told to feck off. I struggled the few hundred yards home, still bleeding and in agony, to be met at the gate by the Da who was furious and wanted to know who had beaten me up. I told him that I would take care of it, for by now the pain had turned to anger, and I was set on revenge, revenge which I had never before sought for anything, nor wanted so badly.

First I grabbed my hurley stick, and then a long leather harness rein. Next, still bleeding, and crying with anger, I went to Rover's stable where I fixed the rein to his collar and led him down the yard, past my Dad. The cornerboys, from where they stood, would be able to see me approaching if I came either from Cork Street through Weavers' Square, or from Brown Street, and would thus be able to make their getaway when they saw my intent. To catch them unawares, I would have to come at them through Hannigan's Lane, as I had so innocently done a short time earlier. To approach this way, I had to make a detour along Donore Avenue, down O'Donovan's Road, back up Susan Terrace, and on through Hannigan's Lane.

As I marched up Cork Street with Rover straining on the long leash, I was priming him for the job ahead with calls of, "Seize them! Attack!"

Rover seemed to know exactly what would be required of him, for he began to growl and show his teeth viciously, and really pull hard on the leash, anxious to get on with the job. People got out of our way as we moved swiftly along the footpath. A policeman on a bicycle gave us suspicious looks, but did nothing. Rover was one hell of a dog, of a size you don't see on the street, and as it was obvious he was in a killer mood, maybe the policeman thought it best not to ask questions. Coming to Hannigan's greengrocery, we moved slowly and quietly, not wanting to scare the cornerboys away before we launched our attack. Rover's hair was bristling, and he growled as we heard the cornerboys give a big cheer, and then laugh over something. With a sudden shout of "Seize them! Attack!" I dashed around the corner, releasing Rover off the long rein and wading into them with my hurley stick.

The attack only lasted a few minutes, but the carnage was great. Rover went straight for the one who had punched me, as if he had been told which one to get first, biting his forearm until he screamed in pain. I went straight for the head of the one who had bent back my arm behind my back, smacking him fair and square on the ear with my hurley. I smacked him again on the shins and arms as he fell to the ground. Rover, meantime, savaged the legs of a third who soon joined his friends on the ground, all three crying out in pain. Five or six cornerboys ran away, but another one with a long pole like a brush handle stupidly tried to beat off Rover who was now wading into the three on the ground. Rover, after getting a couple of blows of the pole, turned on his attacker and bit him viciously on the hand, while I gave him a few clatters on the shins and the back of the head. While the attack lasted, the noise level was very high with Rover's snarling and barking, and the screams and shouts of his victims, and indeed my own crying and shouts of anger. I feel sure that Rover would have killed one of them had my Dad not arrived on the scene and pulled him off by the long rein. Rover then turned on my Dad but luckily I managed to grab his collar before he had the Da on the ground as well.

Three of the savaged cornerboys lay together on the footpath by their beloved corner, howling in pain, while the fourth one crouched directly behind them, leaning against the wall holding his bleeding arm. I stood there shaking and crying, still bleeding and sore from the beating they had given me earlier, and straining to hold back my pal Rover as he snarled, barked, and pulled on the rein to get at them, to finish the job once and for all.

"For God's sake, keep back the dog, he'll kill them," I heard my Dad say. Despite my anger and pain, and my determination to avenge the humiliating and cowardly beating I had been given, I knew in my heart that I didn't want to go so far as killing one of them, and my Dad's call made me realise the seriousness of the situation. By now a crowd had gathered.

Then a policeman appeared from the nearby Newmarket station.

"What happened? Jesus Christ, what has happened here?" he shouted when he saw the carnage.

Just then the father of one of the cornerboys arrived, and after hearing what his bully of a son had to say, made a run at me with something in his hand. But he didn't have to travel far to find trouble, for Rover pulled the rein from my grasp and, travelling at full speed, met him halfway at chest height, knocked him to the ground and bit his upper arm and shoulder viciously, before the Da and I managed to pull him off. Prior to this latest charge, Rover had quietened down somewhat. I too had relaxed a little, not expecting the cornerboy's father to run at me. But in the highly tensed atmosphere, Rover, a very intelligent dog, sensed that the sudden movement towards me was unfriendly, and responded with lightning reflexes. As the man lay on the ground cursing and shouting at me, clutching his bleeding shoulder, another man came out of the crowd and started shouting back at him and his fellow sufferers. He informed the policeman that he had seen what the gang had done to me earlier, and that they deserved all they got. Then two women began to shout at the injured father and the cornerboys, saying that they were nothing but bullies and I should have let the dog finish them off.

A short time later, an ambulance arrived, and all the injured were taken to hospital. One of them had a broken arm.

The crowd was growing by the minute. Another policeman arrived, and, with the first one, started to take statements. While all this was happening, I just stood there, sobbing uncontrollably with my pal Rover by my side. Most people stared at us, whispering to each other. Some crowded around the policemen, telling them that they should have moved the cornerboys on before now, that they did nothing but cause trouble at the corner. After a few minutes I made my way across the square, on towards Ormond Street, and home.

As I moved off, the crowd gave a sudden jump back, opening up a passageway to let Rover and me through. Not even the police tried to stop us, maybe because they knew my Dad, and where we lived.

When Rover and I arrived home, closely followed by my Dad at a respectable distance (because Rover was not on the best of terms with him either), my mother, not knowing what had happened, met us at the door. Seeing my blood-stained face and clothes, and hearing me still sobbing and watching Rover frothing at the mouth, she cried out, "Holy Mother of God, what's happened to you, darling? Who has done this to you?"

She bent down and took Rover and me in her arms, while the Dad, from a safe distance, told her the story as he knew it. I filled her in on the missing parts as best I could between sobs.

I went to my bedroom taking Rover with me, for after such a ferocious battle, the first of its kind we ever had, I felt that we needed each other's company. The Ma and Da followed us up to the room with basins of water, and as the Da cleaned me up, the Ma cleaned up Rover who had got a split lip from a kick in the mouth. After that we were both given a big feed.

Some time after that, two policemen arrived at the door and were brought in to the parlour. After a short time I heard the Da shouting, "No one's going to shoot that dog, no bloody police force or no bloody army in the country is going to shoot him."

Then I heard my normally quiet mother ordering them to leave the house. I heard her telling them several times to get out, and then I heard my Dad shouting at them. But they weren't leaving. At this point I decided to take matters into my own hands, and, grabbing Rover by the long rein which I had re-attached to his collar, I charged down the stairs with the cry, "Seize them, Rover!"

With the speed of Jesse Owens, and the style of the Keystone Cops, they sprinted and half fell out through the open front door, as Rover, even more viciously than in the cornerboy fight, went hell for leather for them, dragging me along with him. Luckily for everyone concerned, the Da was able to grab the rein as we both passed him in the hall, and we were able to stop Rover just inches from the two policemen, who were then lying on the footpath outside our front door, having tripped over each other in their panic to escape.

The next morning there was a letter in our hallway from Dad's police sergeant friend in Newmarket station, asking him to call to see him. My Dad called down to the station. That afternoon both he and the Ma were called down. The next day, both of them went down again, and when they came back they asked me to sit down as they had something important to say to me.

They told me that some people wanted to have Rover put to sleep, and I gathered from their tone that it was not the kind of sleep brought on by Ovaltine. They told me that two policemen would be visiting the house the next morning to see Rover, and it was very important that he should behave himself, for his life depended on the visit.

We immediately set to work to make Rover "policeman friendly" which, in the circumstances, was a tall order. We first had to get my brother Wally out of the way, for we wanted to keep Rover in the house for some time, and with Wally around that would have been impossible, as Rover couldn't stand the look of Wally and Wally, for his part, was terrified of Rover.

When Wally came home from work, we persuaded him to go out to a dance, and to stay out until after midnight. Next we had to make Rover "Da friendly," but as an uneasy truce already existed between them, we thought this would be okay. We got the Da to make up a plate of nice food for Rover, choosing the best pieces from the Clarence swill. As the Da nervously pushed the plate towards Rover, they eyed each other warily.

After a time, Rover got more used to the Da, although you could see from the Da's expression that he was still nervous, and one growl would have blown the truce and sent him diving through the door. Then we washed Rover in the big wooden bath in front of the fire. When we got him out of the bath, just before we started to dry him, he drenched all of us, as well as the walls, the ceiling and everything around, with one massive shake. Eventually he smelled like a bunch of roses and I took him up to bed with me to keep him nice and clean until his interview the next morning.

Rover still smelt sweet the next morning, and, after Wally had gone out to work, we fed him well and pampered him, and I played with him to put him in good humour. Just before 11 o'clock, the Ma put on a fresh cross-over apron, and the Da a new clean collar with his best suit. At about 11.30, a knock came to the door and Dad's policeman friend and another young policeman, both in plain clothes, were admitted by the Da, as the Ma and I cuddled the now very alert and slightly growling Rover. The policemen were brought into the parlour where the Ma, Rover, and me waited. A big fire had been lit, and the table set out with sandwiches and our best china. As the policemen entered the room, they paused and looked at the collarless Rover, while he looked at them. The Ma slipped Rover a piece of meat as the three of us held our breath, hoping it would suffice instead of the policemen.

The younger policeman, obviously an innocent abroad who had not been told all the facts about Rover, suddenly bounded forwarded to kneel before him, holding out his hands and saying, "Here boy, there's a good dog."

My heart missed a beat, when in the next second he placed his hands on either side of Rover's face, and moved his own face to within biting distance of Rover's mouth, as he began to whisper sweet nothings to him. For a second or two Rover gave him that special look, the one he gives when deciding if he should attack or keep his peace. Instantly I pushed a ham sandwich towards Rover's mouth, but the young policeman took it, took a bite out of it, and offered the rest of it to Rover saying, "Good doggie, eat."

In that instant I knew that "good doggie" was about to eat *something*, or *somebody*. To my relief, he opted for the sandwich rather than the policeman. Rover then allowed the policeman to maul him around a bit, and he even licked his hand playfully, and gave him the paw. After that I always thought it a wonder what a good wash, a feed, and a bit of pampering could do for calming you down.

The two policemen had tea and sandwiches with us, and became all friendly-like before leaving. My Dad's friend whispered something to him as he was going out the door, and then, turning back to Rover who was sitting just inside the parlour door, held out his hand and said, "Give me the paw, Rover." And would you believe it, the big slob of a dog gave him the paw and licked his hand for good measure! Had my Dad's friend been one of the two policemen who called a few nights earlier, he would have been chewing it, not licking it. Later I learned that the police sergeant had settled the row on a "knock-for-knock" basis, the cornerboys had dropped their complaint about Rover, and the Da agreed not to summons them for assaulting me.

A few weeks later, when I was halfway down Donnelly's Lane, two of the now recovering cornerboys turned into the lane from the Brown Street end and came towards me. My first inclination was to run, but I decided to brave it out. When we were about twenty yards apart and closing, they came over to my side of the lane, one of them with a plaster-cast on his arm, and the other one limping slightly. For a moment there was silence as we stopped and stared at each other, the eighteen-year-olds, and me.

"You touch me, and the dog'll kill you the next time," I said. Then the one with the plaster-cast moved aside and past me, saying, "Let the little brat go, a tank wouldn't stop that bloody dog of his."

A few months later, when the physical and mental wounds had healed, I again found myself on my own passing the cornerboys at Hannigan's Corner. Just as I was going by, one of them said, "Did you feed that tiger today?"

Then another one said, "Jembo here has another arm, just in case the butcher runs out of meat for him."

Then there was a big laugh from the group. After that, every time I met any of the cornerboys, they would say things like, "Any broken bones today?" "How's Fido?" or "I know a fella with two good arms, do you want one for your dog?"

THE GANG

The kids in our area played with their special friends, or their gang, as they were known. I had an advantage over all of these kids by having the farmyard in which to play, and although on my own quite a lot of the time in the yard, I enjoyed, every minute of it, not bored or unhappy even for a second. But I still spent a fair bit of time with my pals who hung around Marion Villas, Ardee Street, Ormond Street, Brown Street, and The Tenters.

All of my pals were decent kids, from decent hard-working parents, and although we got into various kinds of scrapes from time to time as we were growing up, we were not bad kids in the law-breaking sense, nor could we have been, for our parents would have killed us. But we had the usual adventures like "boxing the fox," getting into fights, going on picnics, going camping, swimming, playing endless hours of football, annoying the life out of the porters of Cork Street Fever Hospital, having running races, bicycle races, and games of cricket and hurling. We played every conceivable street game there was, and when we got the occasional few bob, which was mostly at First Communion, Confirmation, and Christmas, we managed to lose it to each other with our games of "ponner" (pontoon).

The summer evenings were our happiest times, for it was then that we played until we were too tired to stand, or it became too dark for us to see what we were doing.

We were about twelve when we discovered that girls were different from boys, and very pleasantly so. We certainly liked the difference. Our own sisters didn't seem that different from us, being just "one of the boys," so to speak. But other girls, like the beautiful looking and lovely uniformed, sweet-smelling girls from the Holy Faith Convent in The Coombe, now they were something else. Unfortunately, they also caught the

121

attention of the Connolly gang from The Tenters, who were older, smoother, more confident than us, and probably tougher than us as well, if the point ever had to be tested.

The winter evenings, or even the early spring or late autumn evenings, were the best times to chase after the Holy Faith girls, for in the summer we had more important things to do like playing football, fishing, or swimming. Mind you, we did have the odd summer picnic with the girls, and walks along by the canal or through the streets of The Liberties. I was very shy with the girls, and when the gang went down to the Pimlico area hoping to see Julie, Louise, Gráinne, Breeda, and the others, I trailed along behind. When the girls, who met up in Julie's house in St Margaret's Avenue went off walking, I again trailed behind, too shy to join in the *craic*.

One dark winter's evening, when the gang of girls and ourselves were going past The White Swan Laundry in Donore Avenue, we started to play hiding games, darting off here and there in couples to hide. I found myself with Julie, standing close to her, both of us hiding in the shadows of the laundry gateway. As I pressed ever closer to her to avoid being seen, I got the feeling, despite my shyness, that this would be the moment to steal a kiss, for never again was I likely to be so close to her with such a golden opportunity. Shaking and trembling all over, with heart pounding and knees knocking, and fully conscious of the fact that sooner or later a priest would have to be told all about this sinful moment, I gave Julie a quick kiss on the lips. She seemed surprised, but gave me a smile and a little laugh. I felt like Errol Flynn at that moment, a man of the world. I went home walking on air, and spent ages looking at myself in the mirror, trying to figure out if it was the big nose, the floppy ears, or the brilliantined hair, that had made me so irresistible.

Even allowing for the fact that a sex-hungry priest was going to want to know all the details, and I would probably end up saying twenty Rosaries or maybe doing an extra long stint in Purgatory, that kiss was well worth it. But the very next day I nearly died when one of my pals told me that that kiss on Julie's lips would make her have a baby. Up till then, I thought that only married women could have babies.

Then he added, "You'll have to drag the kid around

everywhere with you, and keep wiping his arse and cleaning his nose, and buying him sweets."

Talk about a condemned man's forthcoming execution focusing his mind ... my mind at that time could not have been more focused. I ran home up to my beloved workshop where I panicked in private, and let the tears fall.

In the midst of my depression, Harry Mullery breezed into the workshop, whistling while he looked around for something.

"Are you well?" asked Harry without looking at me as he rummaged around, while I continued to cry quietly. On noticing my tears, he put his arm around my shoulder and asked me what was the matter. Harry, although about twenty-five or so at the time, was a sort of a pal, and I told him, with more tears spurting out, what had happened. With that he burst out laughing. He fairly shrieked, holding his sides and doubling up as he held the side of the work bench for support. The more and louder he laughed, the more and louder I cried, until I just fell into his arms, him laughing and me crying. Harry then told me that Julie wouldn't have a baby.

"Won't she?" I asked.

"Nope," he replied so emphatically that I believed him, and a great cloud passed from over me.

"How do girls have babies then?" was my next question.

"Well," said Harry, thinking as quickly on his feet as a politician meeting his mistress at a Bishops' conference, "all I can say is never blow in a girl's ear."

For a fellow who at that time didn't know the difference between a prostitute and a substitute, I was fair good game for believing anything. When I passed this valuable information on to my pals, I was half laughed at, half believed. For a long time after that, I was fascinated looking at girls' ears, big ones, small ones, round ones, squarish ones, and even the cauliflower ears of a fighting woman from The Coombe. But I was careful to view them from a distance and always breathlessly.

I remember once as a little fellow in Croke Park getting more and more frustrated as a country team was beating the lard out of Dublin. At the top of my voice I shouted out, "Bring on the prostitutes," for I thought that was what you called players who came on for others who weren't playing too good, or were injured. It's amazing how such a call can get the attention of surrounding spectators, for suddenly, nearly everyone seemed to be looking at me. One country fellow with a yellow pullover and no teeth, grinned from ear to ear and nodded approvingly, while another fellow blessed himself and called out,

"Will someone get that foul-mouthed little bastard out of here."

However, the Dublin coach must have heard my shouts, for soon after, the "prostitutes" were brought on and the Dublin team turned the game around.

Julie did not seem to notice me too much after that historic kiss, which didn't say too much for my sex appeal, and my shyness seemed to get worse. It was many a long day before I got another kiss from any girl, except of course from Eileen down the street who followed me around with puckered up lips, but she didn't really count. She certainly didn't make my heart thump or my legs go weak, for she was almost like one of the boys — except of course when we played doctors and nurses. I loved Blue Eyes when she nursed me in hospital after my flying escapades, but Julie was really my first love. Other girls from the Holy Faith in The Coombe who caught the gang's fancy were Fiona from Cromwellsfort Road in Walkinstown, and a friend of hers who cycled in and out with her every day to school.

For a time I found myself fancying Mary Brenton, also from the Holy Faith, who lived just over Dolphin's Barn Bridge. Mary was a lovely shy type of girl, who, as she went past my gate with her friends from school, would give a little glance over to me while I glanced shyly back at her. One day I decided to send her a love note, and one of my pals agreed to deliver it. In my best handwriting I wrote, "My love you're like a red red rose."

The day I sent the note was the same day that Tony Donlon, a friend of ours from Maryland who had gone away to be a jockey, had written home with a tip for a horse. I had written another note, "1/- win Keep Faith."

Unfortunately, Mary was given the wrong note, the note with the bet that should have gone to Kilmartins the bookies. I was so embarrassed at this, so ashamed she found out that I would do such a thing as bet on a horse, even if it was one of Tony's certainties, that after that I couldn't bear to look her in the face, and the flirtation which might have led to full-blooded romance died on the spot.

A few years later I was queuing for a bus in Dolphin's Barn, on my way down to the De Luxe cinema in Camden Street, having failed to get into the Leinster, when along came Mary, directly up to me at the head of the queue with a collection box and flags. As she said, "Hello, would you like to buy a flag?" my heart pounded, my mouth went dry, and I couldn't speak to this Holy Faith vision of loveliness. My hands shook as I put the silver coins into the box, and even the fact that she stuck the pin with the flag into my chest did not disturb me too much, for at that moment she was piercing my heart, as she had done so often in the past with her shy glances. As she smiled and said, "Thank you," before moving off, I was in a near state of collapse, what with the emotion of meeting Mary again, and the pain of the pin. Another story to make the priest happy, I thought.

"Did you touch her, my son?"

"No, Father, she touched me, even made me bleed when she stuck the pin into my heart."

"Did you take pleasure in it, my son?"

"Are you nuts, Father?"

Yes, I could see it all!

After Mary moved off and I got on the next bus, I couldn't think straight, ignoring the conductor collecting fares, who obviously thought I had already bought a ticket. But the bus inspector who got on a few stops later did not allow me to ignore him with his request, "Tickets, please."

When I reached for my money, I hadn't a penny, I had given it all to Mary. I then tried to tell both the conductor and the inspector all about Mary, how I hadn't seen her for years, how in my excitement I just grabbed all my money and put it in her box. I told them how lovely she was. When I had stopped talking, they stood in silence, first looking at me and then at each other. By this time the bus had stopped, and all passenger eyes were upon me. What happened next, happened quickly.

I remember the inspector asking me to stand up, and after that I was propelled at speed by the back of the neck and the arse of the trousers, down along the aisle of the bus and off the platform. As the bus moved off, the inspector stood on the platform, and shouted, "Don't try that trick again."

This love business was beginning to screw up my life.

When Julie was about seventeen or so, she went to work in a Grafton Street store, and sometime after that, she and her family moved out of Pimlico and out of our lives. Shortly after she started work, she told us that her boss's son, aged about twenty-one, had asked her out. Our comment was, "So you're beginning to go out with old men."

I met Julie once again in Grafton Street when we were both about twenty. She had grown from a lovely young girl into a beautiful young lady. We talked, or rather she talked, for in her presence I was spellbound and speechless. As I watched her soft lips move over her ivory white teeth, my heart pounded in the same way it had pounded all those years earlier when we kissed, or rather I kissed, so innocently in the shadow of the gateway of the White Swan Laundry, and earlier when my lips touched her fingers as she gave me a bite of her Mars Bar. Even now, all these years later, a bite of a Mars Bar, with the memories it brings, does a hell of a lot more for me than helping me to "work, rest, and play."

We parted with her call to, "Keep in touch, you know where I am." But although I lifted the phone many times after that meeting to call the Grafton Street store, and stood many times on its threshold, I could not bring myself to take that final step to make contact again with my first love. It was that old inferiority complex again, not helped any by the nose and ears dished out to me by the Da.

Julie and I never met again, but one day, many years later, I was standing talking to someone in Dublin Airport when I heard a lady close by talking and laughing with another lady. It was Julie, I felt sure, but again, shyness, lack of confidence with women, stopped me from approaching her.

———

There was a kid who lived in Chamber Street called Snotty O'Doherty, so named because of the ever-present "candlesticks" between his nose and upper lip, sometimes even dripping down to his lower lip. He was one tough kid, and he had a gang of equally tough kids. Luckily, however, they seemed to play or operate more towards Newmarket or Blackpits, and so we did not have too much contact with them. But the little contact we did have was painful, for, nearly every time we met, they knocked hell out of us.

I was never a good fighter, nor had I any ambitions to be, having seen many horrific adult fights which turned me off violence to a great extent. A friend once gave me a present of a book on Ju-Jitsu, which seemed an easy and effective method of self-defence. Shortly after I finished reading it, two of my pals from Ormond Street, along with myself, were collared by the O'Doherty gang in Ardee Street, held up against the wall and given a few thumps. In a moment of weakness, as I was being thumped, I heard myself saying, "You're very brave when you're with a gang. How brave are you when you're on your own?" I looked around in panic to see where the voice was coming from, but it was mine, all mine. The bloody Ju-Jitsu book by Yang Pan Sing, or whoever, had taken over my mind, but more importantly my mouth. With a look of horror on the faces of my pals, anticipation on the faces of the O'Doherty gang, and feeling somewhat like General Custer on Little Big Horn sending out a call for more Indians, I went around to the waste ground in Weavers' Square, to stand out to O'Doherty.

It was a painful fight — for me, that is. At one stage the bigger, tougher, stronger O'Doherty had my head under his arm and was punching me on the face while I was trying to remember which page in the Ju-Jitsu book, what great defensive trick, would get me out of that situation. Luckily O'Doherty couldn't stand the sight of more than a pint or two of blood — my blood — and in due course he got tired and let me go.

I went home and got Rover, and brought him back to Chamber Street where I set him at the gang. They bolted for O'Doherty's front door, scrambling in just in time, except for Snotty's brother, whose leg Rover managed to bite before he escaped. It was some time after that when the incident involving Rover and the cornerboys of Weavers' Square took place, when Snotty and his gang saw at first-hand the kind of carnage Rover could cause.

A couple of weeks after Rover's massacre of the Weavers' Square cornerboys, I was coming home through Newmarket on my own when I turned a corner and ran slap-bang into Snotty and his gang.

"You touch me and I'll get you with Rover, even if I have to wait outside your door all night," I said to him.

"I'm not goin' near you," said Snotty. "I've seen what that shaggin' monster of a dog did to Nolan and Dempsey," he added. They started to chat about Rover.

"Is it really a dog?" asked one of the gang.

"Its mother was a lion and its father a tiger," said I.

"Wow!" gasped the gang.

Then Snotty gave me a Honeybee toffee sweet and he and his gang went on their way. A week or two later, I was walking along Ardee Street on my way to Meath Street, when I heard the sound of running footsteps behind me, and then Snotty arrived at my side.

"Where are you goin'?"

"To Meath Street," said I.

"I'll be down that far with you," said he. And side by side we walked, chatting away, the past forgotten and forgiven.

I then got another one of those do-it-yourself books, this time on how to hypnotise people, for I desperately wanted to get Brother "Killer" Kelly under my spell. I figured he wouldn't sit on a chair and let me wave a medal on a chain in front of his eyes saying, "You're going to sleep, you're going to sleep." So I produced my ten-conqueror chestnut on a leather thong and swung it back and forth as I sat in my desk facing him. I was hoping that his eyes would make contact with the swinging conker, and that my whispered "Go to sleep!" would also contact his ears in a spell-binding way. Once I had him under my spell, it was my intention to send the hoor to the top of Nelson's Pillar with the instruction to jump. His eyes did indeed make contact with the swinging conker, but unfortunately this happened before the magic, or whatever the power of hypnosis is, had time to work on him.

"Come up here, Boland, and bring that chestnut with you," commanded the almighty one. "What are you up to, and what's all that mumbling about?" was his next question.

And then I did it, I told him what I had been planning, and how he figured in the plan. Even when "Killer" Kelly was beating the lard out of me, saying with each painful belt of his coin-loaded leather strap, "So you'd get me to jump off Nelson's Pillar, would you?" I was nodding my head from side to side between each belt and whispering, "Go to sleep, go to sleep," in the desperate hope that I could still get the bastard under my control, and send him on his way to O'Connell Street, up the long stone spiral staircase of the Pillar, and then down, down, down, to splosh all over the ground.

The next time I tried hypnosis was when a pal and me were going through Pimlico and I tried to help some kids get back their ball which had gone over some railings, on the other side of which an Alsatian dog stood guard. After getting the dog quietened down by swinging my conker in front of his eyes and saying, "Go to sleep, go to sleep," as best I could in a friendly growling kind of voice, which I hoped the dog would understand, I put my hand through the railings to pet him on the head saying, "Good doggie, nice doggie." Suddenly the stupid German son-of-a-bitch bit my hand. After that experience, coming so soon after the Ju-Jitsu book experience, and the "Killer" Kelly larroping, I sort of lost faith in the do-it-yourself books.

We were, in the words of Pete St John's lovely song, "Raised on song and story." My mother would tell me fairy and ghost stories all night, she believing every word and having me believe them as well. My Dad, being a Dubliner, was rather short on fairy stories, as the Little Folk seemed to need a fort under a big oak tree to operate, and oak trees were a little thin on the ground in Cork Street. But he could tell a good ghost story about graveyards in the city, and even throw in the odd banshee story for good measure. Some of the Da's banshees were a little way out, for apart from the normal ones who combed their hair continuously, as all self-respecting banshees do, there was one who played a banjo, and another one who juggled with her head.

These stories frightened me so much that, if the Da told them to me late at night just before bedtime, I had to be escorted up to bed. When I got into bed, I would dive under the bedclothes and sleep like that until the next morning when it was time to get up. Once, after a particularly frightening story, I was shaking with fear under the blankets when Albert the cat, looking for a warm spot to sleep, strolled into the room and made a flying leap on top of me. I nearly lost my mind, and went running, half falling, screaming, out the bedroom door and down the stairs. That night I slept on the couch by the fire, but the next night again, like a turkey going to a Christmas party, I was back for more stories. Many times as my pals and I sat by a warm blazing fire on a cold winter's night in the harness room, I would tell them some of those stories, and then laugh my head off, when at their time to go home, they were afraid to venture out into the dark.

HOPALONG AND MOVITA

The Ma, having come from a farm in Fermanagh, was more used and partial to eating fresh vegetables, and home-made food than most Dubliners. During the war years, things like tea, sugar, butter, bread, and flour were not easy to get, but as far as I can remember, there was no shortage of any kind of meat. From time to time, the Ma took a longing for "country" food, like home-made salted butter, duck eggs, rabbit, goat's milk, and really fresh vegetables. As she had a churn for making her own butter, that part of her yearning was easy to satisfy — that is, if you had the arms of a wrestler, for making butter was not an easy thing to do.

We had a good supply of fresh eggs from our own Rhode Island Red hens, as well as marble-sized eggs from the bantams. We also had chickens to eat, and the Da would occasionally get Mr Williams to kill, cure and portion a pig for us. The farmers who took away the dung, or the manure, as the genteel like to call it, sometimes threw in the odd bag of potatoes, turnips, or cabbage along with their usual payment of straw. All in all, we didn't do too badly. But when we got a lot of vegetables, we gave away most of them to some families in the area who were not too well off.

Cooking could be a problem, for gas was rationed to just a few hours daily. Outside of rationing times, there was still some gas which came through the pipes, but there were gas inspectors going around with the authority to come into your home to check if you were using it. They would put their hands on the gas jets, and if they were warm, even though not switched on at the time, you would be summonsed and fined between 5/- and 10/-, considerable amounts of money in those days. These inspectors became known as "glimmer men."

With petrol in short supply during the war years, some commercial vehicles switched over to using gas, which at the

131

time was made from coal. For a time, this gas was carried on the top of vehicles in a large type of balloon, which sometimes scared the horses. When you consider that this was the same kind of gas that flowed through the pipes, there must have been a big safety risk involved in carrying the stuff around. Certainly you would stop and think before flicking your Woodbine in its direction, or emptying your clay pipe out through a top window. Again, because of the petrol shortage, people had cars which they could not drive. Some enterprising car owners fitted shafts to the cars, and hitched up horses to take their families for a drive.

To give us greater cooking flexibility, the Da and a friend fixed up an ingenious type of sawdust-burning cooker. It was a barrel within a barrel, which gave off terrific heat, and was flame-red when you looked into it. In addition, we had the range fire, and a plentiful supply of coke (coal after the gases have been removed) from Mr Mullery's ration for his bakery business.

My Ma loved milk, particularly goat's milk. Whereas cow's milk was in plentiful supply, goat's milk was hard to come by. But one day my Dad, believing he was going to give the Ma a happy surprise, accepted a goat from a man in lieu of payment of a debt. It was never clear whether the Da told the man what he wanted the goat for, at the time of taking it over. And later the Da refused point blank to discuss it, as Ma and the neighbours fell about the place laughing at him. The day he brought the goat home, I met him at Dolphin's Barn, he driving the horse and cart, and me on my way home from Rialto.

"Get up, I have something to show you," he said, beckoning me over. Climbing up onto the cart, I spotted the goat lying in the back of it, half covered by a sack, its legs tied with rope, and its mouth tied with the Da's neck tie.

"What do you want a goat for?" I asked .

"It's a surprise for your Mammy," he replied.

"Who's going to kill it?" I asked.

"There's no one going to kill it," he shouted back impatiently, then added, "she doesn't want to eat it, only drink its milk."

Then I asked the Da why the goat was muzzled with his tie.

"I only had enough rope to tie its legs together, but after getting it all the way from Firhouse to the Half-way House, I couldn't take any more of its whingeing, with everyone looking up at me, thinking I was the one whingeing. And so I had to use my tie to shut its mouth," replied the Da.

As we came near the yard, the Da, with a big smile on his face, said, "Now here's the plan. I'll back the dray into the hayshed and tip out the goat so your Mammy won't see it. You go into the kitchen and bring out a milk tin. Then we'll draw off a pint or two of milk and surprise her with it. I can't wait to see her face when she sees the great milker we have here."

The Ma was busy when I went into the kitchen, and she didn't see me smuggling out the milk tin. It was made of stainless aluminium, with a lid and a wire handle. We used it every day, as the milk was sold "loose" from the back of a milk float. The milkman lived in Rialto, and had his own cattle on a farm in Bluebell. Everything about the him suggested that he ran a spotlessly clean operation. His horse was beautifully groomed, the harness shone, the float was always clean on the inside, and the outer varnished bodywork and wheels gleamed. The churns were clean and free from any spillage, and the man himself was neatly and cleanly attired in a white coat and white cap. Joe Mullarkey, the icecream maker who stayed in our yard for a time, also had a white coat, but that's where all similarity ended.

By the time I got back out to the hayshed with the milk tin, the Da had unceremoniously unloaded the goat, which landed in a heap. When the horse was bedded down and fed, we turned our attention to the goat. We closed the two half-doors of the hayshed to keep it from getting out, and we untied its legs, but not its mouth, for the Da said its whingeing drove him mad. When we got it up on all fours, I was told to hold its head, while Da set out to draw off a pint or two of the milk that was going to be a pleasant surprise for the Ma.

The scenes which followed could have been taken from a Laurel and Hardy, Buster Keaton, or even a Mickey Mouse picture. Da put the milk tin under the goat, in the same way that you'd put a bucket under a cow, and then he started to milk the goat. But the goat went wild. It wasn't having any of this, and, with a flick of its head, flung me against the wall. Then it turned and ran at the Da, sending him sprawling flat on his back. I was all right but the Da was hurt, and for a minute or two I was frightened, until he slowly sat up. Glaring at the goat, he said, "If it's the last thing I do, I'll get a pint of milk out of you."

With that promise, we took on the goat again, I going for its head once more, and the Da going for the milk tin. But the goat had other ideas, and took off around the shed with the Da and me in pursuit, which was some trick, as the shed wasn't all that big, and was half-filled with hay. At one stage the goat stopped running and charged the Da, hitting him so low that, had he gone to choir practice that night, they would have put him in the soprano section. The Da was really hurting this time, and it took a little longer for him to recover, while it took me all my time to stop laughing, for the neck tie had come loose and was now providing the goat with a tasty snack. The Da at this stage was so hurt and angry, that the Ma's pleasant surprise had lost its appeal.

Red faced, he shouted, "What the hell your mother wants with goat's milk anyway, I'll never know. Me, my father, and his father before him, were reared on honest-to-goodness cow's milk, and look at the kind of man I turned out to be!" That seemed to me an extra good reason to give the goat's milk a chance, but wisely I kept my thoughts to myself. Then lunging at the goat's head, the Da cried out, "There'll be either goat's milk or goat's meat on the table tonight."

No sooner had the Da got his arms around the goat's neck than it gave another terrifying flick of its head, sending the Da crashing against the wall. But like the heroes of old, Da was soon up and at him again, this time aided by Mickey Joe, the dairy boy from a nearby dairy yard, who had heard the rumpus and came in to see what was up. On instructions from the Da, Mickey Joe grabbed the goat's head, while I held its back legs and the Da started the milking again.

"By God, I'll get a pint of milk out of you today, or I'm a Chinaman," shouted the Da.

With that, my mother's laughter interrupted us, she and Bridie Gray having come up to the half-door of the shed without us noticing.

"Put down that goat, Charlie Chan. Whatever you get in that can, I for one am not drinking it," she said.

Then the Da exploded: "I went to a lot of trouble to get this goat. I cancelled a bill worth £4, drove it all the way from

Firhouse whingeing his head off, the little hoor ate my tie, and then it nearly castrated me. I did this all for you, to surprise you, and now you say you won't drink the bloody milk."

By this time the Ma was in hysterics of laughter, but recovering briefly, she whispered to Mickey Joe and Miss Gray, following which the three of them laughed fit to burst, while the Da and I just stood there and the goat finished off the tie.

Then the Ma, composing herself, said, "I didn't say I wouldn't drink goat's milk, I said I wouldn't drink whatever you might get in that can, Charlie Chan."

"Why are you calling me Charlie Chan? That's the second time you did it," said the Da, getting more annoyed.

"I'm calling you Charlie Chan," said the Ma, "because you said you'd get a pint of milk out of the goat, or you're a Chinaman." Then she added, "Don't you know, you Dublin Jackeen, that billy-goats — gentlemen goats to you — don't give milk? And as for nearly castrating you, I don't blame him one bit, for in your search for the milk tap — and I'm sure that was what you were expecting to find down there — you weren't helping his chances of fatherhood very much. I've read about this sort of thing, but I swear to God, I never thought I would ever witness the stupidity first-hand."

The peals of laughter rang out again. The Da just stood smiling wryly, scratching his head, and said, "Holy Mother of God, so that's a billy-goat."

The goat, whose days were numbered as far as living in our yard was concerned, turned out to be quite an amusing animal. He took up residence in the hayshed, the scene of his great triumph, where he had nearly killed me and given the Da a sex change. We called him Hopalong after Hopalong Cassidy.

Hopalong loved to charge at people, especially when their backs were turned to him. He would hide in the hayshed, peeping out from behind the door now and then to see whether any unsuspecting person was standing with his back to the shed. When he saw his luck was in, he would come out quietly, lower his head, and then charge. Men and women alike who stood with their backs to him would have the legs taken from under them, or the arse almost lifted off them, depending on their height and their ability to withstand the impact of the charge. When I discovered Hopalong's penchant for butting, having been the victim of it myself, I set out to get people to stand on the spot where Hopalong performed best. Then I stood back and waited for the fun to start.

Once a man from Cork Street Hospital ran into the yard after me because I threw a crab-apple which hit him on the head. As I hid in the hayshed, I noticed that he was standing right on Hopalong's attack spot with his back to the shed, discussing me with the Da. I opened the shed door and pointed the goat at him. Within seconds, Hopalong, with all the power and passion that such a special occasion deserved, went thundering down the yard, lifted your man right up off the ground, and dumped him on his back. When the injured party picked himself up, he hobbled away as fast as he could, shouting over his shoulder, "You're all fucking mad in this place, even the bloody animals are trained to kill."

I emerged from the hayshed laughing. My Da gave me a smile. Then he gave me a belt on the ears, and told me that if I ever again threw anything at someone, he would wring my yalla neck.

Hopalong continued to give us a lot of fun for a few more weeks, during which time we didn't get many visitors calling, as the news of the goat's peculiar form of greeting spread far and wide. But after he ate the Da's longjohns off the line, the writing was on the wall for him, and, sure enough, one day he disappeared.

My Dad had many pet subjects on which he gave his views, whether you wanted to hear them or not. The subject of rabbits was one. He hated them with a passion. He said that they bred with rats, and how anyone could eat a rabbit was beyond him.

Although we were never short of food, the Ma worked hard to manage her weekly housekeeping allowance, and sometimes, usually because of her generosity to others worse off, she had to improvise to get through the week. The Da liked to take nice sandwiches with him to work, chicken being one of his favourites. The Ma couldn't always afford chicken, but rabbits were cheap, and sometimes even free, if one of the farmers who collected the dung had a rabbit to spare. The Da never knew anything about it, for he would not allow a rabbit across the threshold. But when the Ma's funds were short, the Da was given rabbit in his sandwiches, and rabbit pie for his dinner, and he always believed it was chicken.

Now and then, for a bit of divilment, I would ask the Da, "Did you like your chicken sandwich today?" Or, "Did you like that chicken pie?"

The Da's reply would be, "Ah, sure you can't beat a nice piece of chicken."

When the Ma heard me at this carry-on, she would go alternately pale with fright and red with anger, shaking her fists at me out of his line of vision. But for all the years he was eating rabbit, he never guessed the truth.

My Da liked eggs, fresh "free-range" eggs, as they are now called. But the problem was that as the Ma ran short of money some weeks, she would sell off the odd hen, which reduced the egg supply, and presented us with what may have been the original "chicken and egg" problem.

Ma would catch a hen, pop it in a bag, and give it to me to take it up to a poultry place opposite Donnelly's bacon factory. There it would be weighed live, and I would be given a docket to exchange for money in the office.

The hens usually laid their eggs in the many nests around the yard, but now and then we found eggs in the most unusual places, one being just below the bowl in the outside yard toilet. I kept these toilet eggs carefully to one side for Wally. If he didn't feel like an egg that day, I would save it up until he was in humour for it another time.

Once when Miss Gray complained to my Da about all the dung balls I was throwing at her Union Jack flag, I gave her a present of one of the toilet eggs, just to get my own back. She reacted by giving me a hug, and then in the afternoon she gave me some chocolate. This kept up for a few days before I discovered that the chocolate was Brooklax which was making me shit all the time. After that I went back to feeding Wally again with the toilet eggs, and throwing dung balls at the Union Jack which fluttered from a flag pole in Miss Gray's back yard.

I used to eat a lot of eggs growing up, for no other reason than they were available and free. But one day I went off free-range eggs when I saw a hen pecking a dead mouse. Eggsperts will tell you that hens are not carnivorous, but I know the facts.

If I had to eat an egg, if someone held a gun to my head and said that my life depended on me eating one, it would have to be a factory mass-produced egg from a hen kept away from little furry animals and fed on corn and such-like, but certainly not a free-range egg. The same goes for milk. Give me pasteurised milk any day, instead of the farm-fresh type you can still get down the country.

We bred our own chickens, having a constant supply of clucking hens, thanks to the efforts of the cock, Errol, whose enthusiasm for his work and dedication to the needs of his female flock would be the envy of many a Leeson Street politician. When a hen was clocking, the Ma would buy a dozen specially selected hatching eggs. The hen would sit on them for three weeks, until the baby chicks broke through the shell. It was a lovely sight to see the hen walking proudly around the yard with all her little yellow chicks darting around her, chirp, chirp, chirping all over the place.

Once I put a duck egg under a hen for a laugh, and it nearly gave her a mental breakdown, for it kept her on the nest for an extra week, so she couldn't be with her chicks which went running and chirping all over the place. When the duckling finally hatched, it had to be hand-reared, for neither the mother hen nor the other inhabitants of the yard wanted anything to do with it. The Da thought the incident funny but the Ma, who had to rear it, was not amused.

The other thing I remember about eggs is that they used to be code-marked each week for freshness, the code number being read out on the radio, after the *News*. It was the responsibility of the egg suppliers to stamp the code-mark numbers on to the individual eggs, using a rubber stamp and pad. That system of advising users how to identify fresh eggs continued right up until the 1970s, when EEC regulations required that such information be shown on the containers. Not knowing as a kid how the marking was done, for a long time I marvelled at the cleverness of the hens who could print the code-mark number on the eggs each week before they were laid.

For a time the Ma went off bought milk. She wanted to have her own cow, and get her milk straight from the source. Wisely, the Da decided to borrow a cow from a nearby dairy, just to try out the idea, before committing himself. Mickey Joe, who worked in the yard, offered to bring us a bucket of milk each day, saying it would be stupid for us to borrow a cow, as we would have to mind it and milk it ourselves. When the Ma declined this offer, he said he would come into our yard and milk the cow for us. This offer was also declined, for reasons which even his best friend would not tell him.

In fact, the Ma was loath to have anything whatsoever to do with any beast that Mickey Joe might have handled. But she was persuaded by the Da that the cow should not be held responsible for Mickey Joe's lack of hygiene or style of living. In due course a cow arrived. The first thing that happened was that the Ma gave it the mother and father of a bath, although not in the wooden tub in front of the fire. She paid particular attention to the udder and the region of the tail. When she had finished washing it, all it needed was a red ribbon and it would have been the hit of the Garda Ball up in *The National* in Parnell Square.

The Ma drew off the first few pints of milk and gave it to the pigs with appropriate apologies, just in case Mickey Joe's influence extended further than it should. After that, remembering our valiant attempts to get milk from the billygoat, the Ma showed me and the Da which bit of the cow to pull.

"Where's the tap?" asked the Da, grinning at me.

"It used to be connected to your brain, but someone léft it running, and that's the reason you are as you are, Charlie Chan," came the Ma's reply.

The experiment worked for a week or two, the Ma claiming that the milk was "pure cream," and it looked for a time as if the cow would move in permanently. Because of her big brown eyes, we named her "Movita," after the beautiful Mexican wife of the famous Irish boxer Jack Doyle. Movita, with her exotic name, and her lovely creamy milk that we couldn't get enough of, brought a little bit of romance into our yard.

But one day the bould Mickey Joe blew it. He was sneaking down the yard with a bucket of Movita's milk when the Ma came out and caught him.

"Just a half-bucket to top up a churn," was Mickey Joe's excuse.

It was not that the Ma begrudged him the milk, for after all, the cow had not yet been paid for. It was just that the spotless Movita had teats so clean, prior to Mickey Joe getting his hands on them, that you could drink straight from them. The speculation as to the many places those unwashed hands could have been, and the likelihood of them getting at the teats again, all proved too much for the Ma, and so Movita was returned to the dairy yard, no doubt to be queen of the herd.

The Ma's last words on the subject were, "With that dirty oul' fella coming in and out at will, even a bull elephant wouldn't be safe from some kind of disease."

STARTING TO SMOKE

I started to smoke when I was about four years old, and continued until I was eighteen, when one day one of my bosses at work told me to give up smoking, at the same time announcing that he was going to train me for athletics, after I had won a half-mile race in the company sports. As I had had a lot of difficulty getting the job in the first place, there was no way I was going to cause his displeasure, and so I gave up the evil weed on the spot, along with the bottle of stout which I had started taking with my pals on a Saturday night prior to going to a dance. One of my pals, a very good athlete with great potential, continued smoking and died at a comparatively young age from lung cancer. Others I knew, and even drank with, for that short time, became alcoholics and brought nothing but misery to themselves and their families. So all in all, at a young age, I managed to eliminate the "cigarettes and whiskey" from my life, leaving only the "wild, wild, women," which unfortunately I didn't meet enough of, that is if you exclude the girls from the Holy Faith Convent in The Coombe.

At about four years of age, my pals and I used to gather cigarette ends off the streets and bring them back to our yard where we opened them up and heaped all the tobacco together. The cigarettes end, or "butts" as we called them, were not that easily come by, because cigarettes at $2^1/_2$ old pence for five Player Weights or Woodbines were expensive to the poor or unemployed, who either smoked them until there was nothing left, or saved the butts to collect the tobacco and roll new cigarettes on little rolling machines, into which cigarette paper was inserted. We rolled our own in brown paper or simply packed the recycled tobacco into cheap clay pipes and lit up. Sometimes we sat around in a circle, smoking and passing around our pipe of peace, pretending we were chiefs from the great Sioux nation, and discussing what we should do about the pale-faced Snotty O'Doherty gang from Chamber Street. Whatever chance O'Doherty and his gang had of getting a quick death, not so the Christian Brothers, and the many ideas of slow torture devised for them at those gatherings were just brilliant.

Sometimes we mixed turf mould with the tobacco, whenever the harvest of butts was not very good, or when there was a high ratio of pipes to tobacco. On occasions we bought the odd cigarette which the shops would sell separately out of the packet, and this we would share out, puffing it between us. Occasionally there was the odd "contribution" from a parent. We also made cigars by rolling the collected butts inside the brown paper. Cigar smoking was dangerous, because we could never manage to compact the mixture very tightly, so, if it burned at all, you ran the risk of sucking a loose mass of fast-burning particles into your mouth. This happened to us often and even though we got our throats burned, we still managed to proclaim in no uncertain manner how upset we felt.

These smoking sessions were usually held in our hayshed, the most comfortable, though not the safest of places. Once while on my own, I dozed off in the middle of a smoke, letting the lighted pipe fall out of my hand onto the hay around me. I was awakened by a bucket of water thrown over me by Mickey Joe, who had spotted the flames as he was passing the shed. Another time, the Ma, on hearing laughter and shouting coming from the shed, suddenly opened the half-door, almost catching a gang of us in the middle of a great smoke. The only way to hide the pipes was to sit on them. As we sat there smouldering, with the smell of "ham" wafting through the smoke-filled shed, we had a hell of a job convincing her that it was steam rising from the most recent delivery of hay which had not been properly dried.

I was nearly in big trouble when one fussy horse suddenly gave up eating the hay, and the Da, trying to find out the reason, got a smell of the smoke and promptly condemned the poor farmers. "Those lazy farmers, all they have to do is three months work in the year, and they can't even do that right without smoking in the haycocks and destroying the hay."

After that we abandoned the hayshed for our smoking sessions. I felt the Ma and Da knew I smoked, and although they had questioned me a lot over the years, they couldn't get enough evidence for a conviction, which is as well, for the Da would have killed me. Years later, at the age of eighteen when I had given up smoking and was beginning to get involved in athletics, I was packing my bags one day to go off training when the Da said, "I'm glad you gave up the cigarettes, son, you're looking a lot better, and I'm sure you'll run a lot faster." You could have knocked me down with a feather, for I never thought he knew that I smoked.

"Who said I ever smoked?" said I.

"Don't act the maggot, you've been smoking since you were knee-high to a grasshopper, and it was just that you were such a clever little devil, that I couldn't catch you at it," said he. Then he added, "I suppose we were lucky you didn't burn the place down, or for that matter kill yourself with some of the concoctions you smoked."

The Ma, who was sitting in the corner darning socks, then smiled at me and said, "Steam from the damp hay, you told me one day, when the place looked like the Great Fire of London was burning. Did you gang of toughs think I came up on the last load?"

Surprised, to say the least, I just flopped down onto the couch and said, "Why didn't you say something ?"

The Da said, "I didn't have enough proof when you were young, and I had to have proof before knocking the lard out of you."

After a few moments I smiled and said, "I bet you never had this trouble with the other fellow?" Then we all laughed.

One day, "the gent", a poverty-stricken old gentleman who had obviously seen better days, strolled past our gate with a dirty big cigar in his mouth. He wore a carnation in his dirty lapel, a cream-coloured straw boater on his head, white spats over his worn shoes, trousers so highly half-masted that half of Dublin must have died, carrying an umbrella in the crook of his arm — altogether a splendid sight, much more pleasantly observed from an upwind position. But what caught our interest was the cigar. We hardened seven-year-old smokers took one look at that cigar, and, imagining the contribution such a butt would make to our tobacco collection, went after him calling out, "Butts on you," to let him know that we wanted the end of his cigar.

We followed him up Cork Street, over Dolphin's Barn Bridge, then beyond Loreto Convent at Sundrive Road, as he strutted along smoking his beloved cigar, which by now was also our beloved cigar. Then we went on up the Crumlin Road, past Iveagh Grounds, now almost walking beside him, trying to find out when he intended throwing away the butt. Our enquiry, as we reached the top of the Crumlin Road, was brushed aside with a wave of his hand holding a progressively smaller cigar, with the comment, "Get away from me, you dirty little boys."

Talk about the kettle calling the pot black! When we reached the Half-Way House, miles from where we lived, didn't the louser stub out the cigar with his fingers and put the butt into his dirty waistcoat pocket. Then as a 50B bus going back into town stopped beside him, he jumped on it, leaving us penniless and tired to walk all the way back without even the butt. Thereafter we changed his name from '"the gent" to "old bollicky."

WALLY AND THE SOW

Wally, my half brother, was about eighteen years old when I arrived in this world. He was the apple of my father's eye, for it seemed that throughout his life, he never did anything wrong, nor did he say anything wrong. He went to work in the Corporation at an early age, progressing from messenger to the pick and shovel, and then on to lorry driving, and other work later in life. The day Wally became a lorry driver, a degree from Trinity could not have made my Dad any happier or more proud of him. As far back as I can remember, my Dad and Wally would sit by the fire almost every night after work, discussing the events of the day in the Corporation. Wally was a very serious and conscientious man, tense at times, but a good man. He seldom smiled, and I cannot remember ever hearing him laugh.

Throughout my childhood, as I embarrassed and enraged my father with one act of devilment after another, and incited him many times to chase me down the yard trying to get his hands around my neck, I could hear his war cry, a cry that was with me all my growing up years, "You'll never be like the other fellow."

"The other fellow" of course was Wally, the perfect one. If "the other fellow" had been born with a halo and been able to lay gold bars, the Da could not have loved him or admired him any more than he did. In short, Wally was one hell of an act to have to follow, one which, early on, I decided I wasn't even going to try to follow, for the same reasons that I left the big mountain to Hillary, and the long boat ride to Chichester.

Wally was always a little tense, and sometimes nervous when doing jobs. One job which made him both tense and nervous, and even downright afraid, was when he stayed up all night to look after a sow having pigs. We kept one stable particularly clean and well-lit for special occasions, such as sows having pigs, sick animals being treated, or horses being shod.

One of the dangers for newly-born little pigs is that of the mammy pig throwing herself down on top of them, killing them in the fall. Sows are inclined to position themselves close to a wall when they want to lie down or throw themselves down, so a baby pig trapped in such circumstances has little chance of survival.

To counter this danger, the Da had special wall creels built into the ground around each wall. These were wooden posts about four inches square, sunk in concrete, protruding about twelve inches above the ground, and eighteen inches from the wall. To these posts he bolted on wooden beams of similar dimensions, which ran right around the four sides of the stable. This meant that the sow could only lie down or throw herself down against the beams, thus giving any little pig likely to be trapped the opportunity of escaping out underneath the beams. The support posts were not sunk directly into the concrete, but inserted into a metal 4-inch sleeve which had been sunk in the concrete, thereby enabling the creels to be removed when the stable was needed for another purpose. All state of the art stuff.

Sows farrowing — or "pigging" as the pig men would say — was a regular occurrence in our yard and one which I looked forward to, because it meant getting a day off school. It meant other things as well, such as sitting up all night in the pig sty, looking at the sow while it looked at me, and we both grunted at each other from time to time.

Then there was the picnic. At about 10 pm my mother made up a flask of tea with sandwiches, biscuits and cake, to help me through the night. Not quite what Kris Kristofferson would need … Armed with my food parcel, a bag of sweets, and a heap of comics to read, I would take up my position on a comfortable armchair close to the sow as she lay by the wall creel.

Outside it could be dark, sometimes with the wind blowing and the rain lashing down off the galvanised roof, shaking the sheeting and making an almighty din. Then as the witching hour of midnight arrived, when most civilised people would be home tucked up in their beds, human noises from nearby houses and passing street traffic would be replaced by animal

noises from the adjoining stables and sties. Sitting there, in my
comfortable armchair, all tucked up in my Foxford rug, a mug
of tea in one hand, a big lump of Ma's apple cake in the other,
a mammy pig grunting contentedly at my feet, a paraffin oil
heater keeping me as snug as a bug in a rug, and content in the
knowledge that I wouldn't have to face "Killer" Kelly the next
day in school, sure life couldn't have been more perfect at that
moment. What more in the world could a young fellow of my
age have asked for?

As the night wore on, and the sow and I stole meaningful
glances at each other from time to time, and I read my *Beano*,
Dandy, *Champion*, *Hotspur*, *Radio Fun*, and *Film Fun* comics, the
low background noises of the other animals in the yard became
familiar and anticipated. But occasionally some sudden and
unfamiliar noises would be heard, and coming in the middle of
a black silent night, maybe just as a sudden gust of wind would
blow open one of the half doors on the stable, it could sure play
havoc with the imagination if you let it. Could it be one of those
giant spiders I had just read about? Or the ghost of someone
who had just died in nearby Cork Street Hospital, risen up out
of the mortuary and making its way over the roofs and into our
yard? Or maybe …? Or maybe …? And so it went on, all the
time tingling my spine while my hair stood on end. Going
through life I learned that it was the living one had to fear, not
the dead. As for the hair standing on end, I wish to God I had
hair today that *could* stand on end.

Once in the middle of the night, as I sat reading all about
"Rockfist Rogan" in one of my comics, the Ma arrived silently
in her long white dressing gown, and stood there just inside the
door without saying a word. I had been dozing in the chair and
did not hear her come in.

When she whispered, "Is she nearly there?" and I turned
around, half asleep, to see the lady in white with her white face
standing there, I let out a scream thinking she was a banshee
with a bit of Kitty the Hare thrown in for good measure. I
collapsed off the chair, and at the same time did quite a big
bowel movement. After that I understood why people who
stayed in haunted houses at night wore brown trousers. My
reaction sent the sow into a frenzy. Never having heard of the
banshee or Kitty the Hare, or the oul' wan who it was said

sometimes danced on the mortuary roof of Cork Street Hospital with her head in a brown paper bag, the terrified animal went straight for the Ma. Like a shot, the Ma screamed and ran out the door for the safety of the hayshed, and she got the door closed and bolted just in time, before the sow clattered into it.

All this noise at 3 o'clock in the morning brought things to life, for the horses started to whinny, the pigs grunted, the dogs barked, the cows in the nearby dairy mooed, and the Da arrived in his longjohns. Mickey Joe, the dairy boy who slept with the cows in the dairy yard, also arrived, having climbed over Miss Gray's garden wall to get to us. All around, heads appeared at back bedroom windows as the neighbours wondered what was up.

My Da, always a man for getting his priorities right, got the sow back into the stable and calmed her down, as I cowered sticky and uncomfortable in the chair. Then he rescued the Ma and got her quietened down and into the house. Next he got me inside too, and after I was calmed and cleaned up, asked me what happened.

Before I could answer, the Ma said, "Your son let a scream out of him for no reason at all and nearly frightened the life out of me, and drove the sow wild, which then tried to kill me."

"Your son," I managed to stammer out to the Ma, for I was still in a state of shock, "was sitting there quietly minding his own business and reading a book, when you sneaked in, all dressed in white and whispered 'Is she nearly there?' I nearly died with fright, I thought you were a banshee."

With that the Da took a fit of laughing, and after a few more arguing words, we all saw the funny side of it, and had a good laugh together.

The Da decided I had had enough excitement for one night, and that Wally, who had slept through the whole shenanigans, should get up out of his warm bed and replace me. Wally was not pleased.

"Bloody sissy, afraid of your shadow," he jeered as he made his way out to the stable. When he called me a sissy, something clicked, and I knew that one day I would get my own back on him. The next morning I awoke early, but stayed quiet until I knew it was past school-going time. When I heard the Ma's footsteps, I pretended to be asleep. She came into my room and stood beside me, looking down at her own little darling thinking how lucky she was to have me.

"I could have killed you last night," she whispered, then kissed me and left the room.

The sow pigged the next evening, just as the Da arrived home from work. I was in the middle of things, cutting the cord and cleaning the little pigs as they arrived — fifteen of them. It was a good litter. Little pigs arrive with needle sharp teeth, and if the tops are not snipped off, they will hurts the sow's nipples when sucking her. Sows have been known to eat their young, and receiving painful nipple bites could be one good reason for doing it. My Da held each little pig on his lap with a piece of wood pressed back on its jaws to prise open its mouth, while with the other hand he clipped off the tips of its teeth with a small sharp wire cutters.

Time moved on a few more weeks or months, and another sow was due to have pigs. Da felt that the coming Thursday night would be the one to stay up. And so, with the memory of the banshee behind me, I started swopping my comics and books so that I would have plenty to read. I also started saving for the odd bar of chocolate or bag of sweets to help me through the night. The Ma, realising that this is my come-back night after the banshee incident, made me some of my favourite corned beef sandwiches, and promised not to frighten me. Harry Mullery, anxious to support his old pal, gave me a big bag of home-made toffee, saying his sister Annie will never miss it.

At half-nine on the Thursday night, just as I had taken the armchair out to the stable and was gathering up all my bits and pieces, including the flask and food parcel, Wally arrived home to announce that he would stay up with the sow, and I could go to bed so I could get up for school the next morning. I pleaded with Wally to be allowed to stay up, even offering to share the watch with him, so desperate was I to stay away from school.

But Wally rejected the offer, Albert Reynolds style, with a sweep of his hand, saying, "You're a big sissy."

I then appealed to the Da, but he sided with the perfect one, saying, "Wally is right, you should go to bed and to school tomorrow."

I lost the argument, but flatly refused to hand over the flask and food parcel, reaching for my hurley to emphasise my determination to salvage some modicum of pride from the encounter.

Disappointed, I went to bed, but with my picnic intact. In order to help cheer me up, the Ma carried a fire up to the fireplace in my bedroom, and heaped it up with enough coal to burn most of the night. We very often carried coal fires from one room to another, as it was a great way of getting a fire started quickly. A few blazing coals were taken from an existing fire and carried in a metal shovel or in a metal bucket, to become the firelighter for a new fire. A fire carried like this into a cold room meant instant heat.

It was about midnight, as I sat by the flickering fire with its shadows dancing on the wall, when I got the idea to screw Wally.

How was I going to do it, and more importantly, how was I going to survive afterwards, were the most pressing questions. It was nearly 1 o'clock in the morning when I finally worked out the plan.

I started putting it into action by first getting dressed, putting on my runners, and then sneaking down to the stable where Rover was housed. Normally when all the stables were locked up at night and the yard gate bolted, Rover would be let loose as a watchdog. And I can tell you that any burglar attempting to steal from our yard would be one sorry man if Rover got him. Rover was my pal and I could wrestle him and even ride him around the place. My mother, who fed him when I wasn't around, could also do anything she liked with him.

But Rover did not like my Dad all that much, for some reason, and he absolutely hated Wally. Every time he saw him, he went mad. Wally, whose nervous disposition didn't help any, was terrified of him.

The problem between them arose when Rover was a pup, shortly after he arrived in the yard. One day Wally gave him a kick to get him out of his way, and Rover went for him, chasing him into an outside toilet in the yard, and repeatedly throwing himself at the toilet door, trying to get in at Wally. Rover never forgot that kick, and Wally never forgot Rover's anger. So, when Rover was out in the yard on watchdog duty, Wally stayed well clear of him. When Wally was out in the yard on his duties, Rover was either locked up, or kept on a lead. A few times, however, when Rover was lying contentedly in the hayshed, Wally ventured out into the yard, being very careful to stay at a safe distance. On this night, while Wally baby-sat the sow with the stable door unlocked, Rover was safely locked up.

Although Rover was a powerful and vicious watchdog at night, he was pretty quiet during the day and could be freed to wander around the yard. If taken outside of the yard, however, he had to be on a lead, for if anyone teased me, or attacked me, or even looked sideways at me, he would go for them. Once in the yard, Wally grabbed me by the back of the neck and gave me a belt for some devilment I had been up to, not realising that Rover was loose in the open hayshed and watching the incident. With a bark that was more like a bellow, Rover bounded down the yard to get Wally, who luckily was standing by the harness room door at the time, and was able to make his escape. As Rover threw himself at the door, Wally stupidly came to the window, which was about five feet off the ground, to tell me what he would do with me the next time he got me without the dog.

Like a flash, Rover was through the open window, knocking Wally sideways. Howling in terror, Wally managed to escape somehow. His howls were almost drowned out by the deafening barks of my brave protector. Outside in the yard, I nearly wet myself laughing. The message was loud and clear — no one messed with me when Rover was around.

But back to Wally and the sow. Quietly I sneaked past the shed door where I peeped in to see the two of them sitting staring at each other. I crept into the stable where Rover was kept and led him out by the collar, silently past the shed with Wally and the sow. I took him on up to my bedroom, and settled him happily on the big double bed. I was now ready to put the next part of my plan into action.

The stables and the pig sties each had two half-doors which could be bolted separately. Only the bottom half-doors would be bolted during the day, allowing the horses to look out, and the stables and sties to be aired. The two half-doors would be bolted at night. But when anyone was in the shed, the two half-doors were just closed over, not bolted, for safety reasons. A sow about to farrow can be a dangerous animal. I have heard stories of sows killing people, breaking legs, and causing other injuries. Luckily none of us ever suffered injuries from a sow.

With Rover asleep on the bed, I sneaked downstairs, through the kitchen, into the harness room, and then out into the yard. Before going into the yard, I pushed over the bolt on the inside of the harness room door, and locked it into position with a padlock, putting the key in my pocket and not on the hook under the shelf where it was usually left hanging. Then, having secured the door, I climbed through the window out into the yard, and on to Wally and the sow.

Very slowly, and without making a sound, I pushed over the bolts on both halves of the stable doors, locking Wally in with the sow. Next I helped myself to a long-handled shovel, which I gripped firmly near the end of the handle. With the shovel held over my shoulder, and my legs braced firmly apart, I stood just a few feet from the iron-clad stable door, behind which Wally and the pig were imprisoned. Then, imagining that I was the big fellow hitting the gong at the start of a J. Arthur Rank film, and at the ungodly hour of half-one in the morning, I started to get my revenge on Wally for calling me a sissy a few weeks earlier when I had shit my trousers with fright.

I banged the shovel onto the heavy door as hard and as fast as I could, at the same time giving the Tarzan yodel at the top of my voice. The response from inside the stable exceeded all my expectations. Wally, who I said earlier was intense and sometimes nervous when doing certain jobs, was in both of those states when minding the sow, for he knew he was dealing with a potential killer, a regular Christian Brother of the animal yard.

His first reaction to the racket was to try to escape through the door. On finding it bolted, he started throwing himself bodily against it. The sow, startled by the rumpus, went bald-headed for Wally, screeching and grunting in her panic. Wally, for his part, shouted for help, and screamed in unison with the sow.

She chased Wally around the stable, crashing into his chair, against the creels. He gave the occasional push at the door in passing, in the forlorn hope that it would open. I kept banging the shovel harder and faster, yodelling all the time. I was enjoying myself thoroughly.

By this time, the neighbourhood had woken up at the commotion coming from the stable. I dropped the shovel, ran down the yard, and jumped in through the harness room window. As I scuttled into the house, I met the Da running out in his bare feet and long-johns, yelling, "What in the name of God is happening to poor Wally?"

I met the Ma as I ran up the stairs, but her only comment was with regard to my safety:

"You've done something wrong, I can feel it in my bones, and they'll kill you for it."

"They'll not lay a hand on me," I replied over my shoulder.

When the Da reached the bolted harness room door, he couldn't find the key to open the padlock. While he flootered around looking for it, both performers in the stable were getting even more excited and dedicated to their tasks at hand, judging by the ever-increasing noise level.

The Da couldn't figure out what was going on inside the stable, except that his beloved perfect son was in danger. With the strength born of panic and paternal love, he launched himself at the window sill which was a good five feet off the ground. He gripped it with his vice-like hands, and with the Ma lifting and pushing as best she could from behind, he managed to drag himself onto the ledge, and beyond to the yard. But the Ma must have pushed a bit too enthusiastically because he toppled over and landed in the yard like a sack of coal. Though winded and hurt, he drew inspiration from his hero, the brave General Rommel, who almost single-handedly kept back Montgomery's 8th Army of Desert Rats. Picking himself up, he pushed on, muttering that he would choke whoever was responsible for starting all this trouble. But the Ma made no comment, for she guessed that I was responsible for it and for Wally's plight, whatever it was. In the dying moments of the drama at the window, I considered it prudent to sneak down the stairs and leave back the padlock key, doing so just as my Dad belly-flopped out through the window into the yard. As I withdrew to a safe vantage point from which to watch the rescue, the Ma shouted after me, "You're for the high jump," to which I confidently replied, "I'm not!"

When Wally was rescued, he was not in the best of shape, physically or mentally. He had never been a sportsman and was far from fit, but that night, in a very confined space, he ran further and faster, and jumped higher and more often, than he had ever done before in his life, or was ever likely to do again. Throw in the chasing killer pig for good measure, and there was an experience he could dine out on for the rest of his days, and would remember on his death bed — a piece of furniture he was fast making his way to when rescued. Fair play to him, though, for he went in and out and over the wall creels so fast that there was talk for a time of entering him for the Irish Grand National.

When the Da finally released Wally, he staggered out into the yard, with the clothes stuck to his back, and the sweat just pouring off him. He was shaking all over, and his head was twitching from side to side, as he and the longjohn-ed Da helped each other back to the house. They had to go the long way around, as the harness room door was still locked. By the time the injured parties reached the front door, I was up in bed with Rover.

I heard Wally shouting, "I'll kill the little bastard," and the Da adding, "I'll kill him too."

The Ma was shouting equally hard: "You'll not lay a hand on the child."

But now fuelled with a killer instinct rivalling that of the sow, Wally headed up the stairs to get me, closely followed by the Da, equally motivated. They were in for a shock.

The first element of my defensive plan was to take out the light bulb, so as to confuse my assailants in the dark. The second element was Rover.

Wally was the first into the room, shouting, "Wait 'till I get my hands on you," as he first tried to switch on the light, before diving across the room with hands outstretched reaching for my neck. He was halfway to the bed when he heard the ear-shattering bark and saw the white razor-sharp teeth glinting in the firelight, as the monstrous Rover rose up like lightning from the bed and went to meet him. Although holding Rover firmly by the collar, I was still dragged unceremoniously off the bed and across the floor towards Wally, as now, once again screaming and terrified, he tried to make his second escape of the evening from a killer animal.

Backing up quickly, he hit the oncoming Da, who had been travelling like the Flying Scotsman before being derailed. Wally ran, scrambled and fell backwards out through the bedroom door in his efforts to escape, but not without first being bitten on the hand and arm. The Da, who had performed so magnificently when he scaled the five-foot window and belly-flopped to rescue his favourite son, now had some lovely teeth marks on a big toe to add to his injuries. Rover, snarling and snapping viciously, was intent on following the perfect ones down the stairs, and disposing of them once and for all.

As Wally and the Da scrambled out on to the landing, Rover went after them. I was dragged along, still clutching his collar. He would have got them too, if the Ma hadn't been on the landing and grabbed Rover's collar with me as we hurtled past. Between us we got Rover back into the bedroom and quietened him down. Downstairs the Da and the perfect one had barricaded themselves into the kitchen and were licking each other's wounds, cursing me and Rover.

A little later, as I lay in bed with Rover, my mother came laughing up the stairs and into my room.

"Are you awake?" she whispered.

"What is it, Ma?" said I. Rover sat bolt upright in the bed as if expecting another eejity attack from the perfect ones.

"By God, I'm proud of you," said the Ma, adding, "if the men of 1916 had had a brain like yours, they'd have whipped the cowardly English." Then she gave Rover and me a big hug and told me that the Da and Wally would be sleeping downstairs that night. I had one final thing to do which I couldn't resist. I went downstairs and called in through the door:

"Wally!"

When he replied, "What do you want, you little cur?" I shouted back:

"Bloody sissy, afraid of your own shadow."

The next morning I learned that the Sow had fourteen pigs during the night without anyone's assistance, and that the bowels of the perfect one had given up the ghost the previous night when confronted with Rover in full flight. For a few weeks after that, I felt it desirable to keep Rover with me whenever the Da and Wally were around.

In time the Da's injuries healed, and he laughed at the whole episode, and ribbed Wally about how he had been out-manoeuvred. Wally was not amused for quite a long time after that, but he did get over it. When Wally got married, it was to a lovely cheerful girl named Doreen, and when, many years later, I told her about the incident, she laughed her head off.

THE BLACK STALLION

All the animals we owned were well taken care of. If they became unwell and my Dad could not cure them with his range of medicines and rubs, a veterinary surgeon was called. Horses are clean, intelligent, kind animals, and I loved being around them. The dogs, and to a lesser extent the cats, were likeable too, but it was impossible to generate any feelings of love for the poor pigs or hens. Maybe it was because we knew they were only passing through, on their way to someone's table.

The death of a loved animal, as happened from time to time, was a moment of great sadness for us. Once, one of our cart horses was severely injured and had to be put down, after a wall collapsed on top of it. Another horse, given on loan to a relation, was so badly mistreated and starved by him that all our efforts to revive it failed. From the time that horse died, my Dad never again spoke to that relation. Sadly, I was responsible for the death of a cat, something I have never forgotten. The cat had gone asleep under the wheel of a cart, and not noticing it, I drove the horse and cart forward, killing the cat instantly.

But the death of one special horse hit us particularly hard, mainly because of the way it died. When you run a farmyard, the death of an animal is something you have to expect and accept, but this one was different. The horse was a beautiful black stallion, with a white blaze on its forehead. The Da bought it from a man while it was still out on grass on his farm in Crumlin. A week or two later, when the Da got it home to our yard, we discovered that it was unwell. For the next week, aided by the vet, we did all in our power to cure it, but to no avail. This beautiful kind animal, this gentle creature with the big soft eyes, lay on the floor of the stable, unable to stand. Eventually it was decided that, as it could not be saved, it would be more humane to have it put out of its misery.

The man from O'Keeffe's the Knackers was called, and he

drove his heavy horse, pulling a deep-sided low-backed float into our yard, backing it right up to the stable door where our beautiful horse lay in pain. My Dad was terribly upset at what was about to happen. Though I dare say he had many times in his life witnessed the shooting of horses, being a horse-lover first and foremost, it was going to be a very painful experience for him. Dad told me to go into the house and stay there, but I hid in the hayshed, from where I could see into the stable. The man from O'Keeffe's produced a gun and stepped towards the prostrate animal, which suddenly raised its beautiful head to look him straight in the eyes, as much as to say, "Get on with it." The man placed the barrel of the gun tight to the middle of the stallion's forehead, exactly in the centre of its white blaze, and pulled the trigger. Immediately there was a muffled bang, the stallion's head jerked, and hung in the air momentarily, before crashing down on to the stable floor, as blood spurted from what was once a lovely white blaze on a shiny black elegant head.

The gun used was probably a captive bolt pistol which, together with an instrument called a pitting rod, was referred to at the time as a "humane killer." The gun fired a self-retrieving hollow bar into the animal with great force. After the gun was fired, the pitting rod would have been used to complete the execution, in a way I cannot bring myself to even think about, let alone describe. But these days, thankfully, there is no question as to the humaneness of such killings, for horses die instantly from special free-flying bullets, fired from powerful hand-held pistols.

The memories surrounding that execution have stayed with me all the years, and if I think about it, alone in the quietness of the night, I relive the numb terror of the moment, smell the stable, and taste the tears rolling down my cheeks.

I stood mesmerised in the hayshed as the gruesome episode was played out. The man from O'Keeffe's, an expert at his job, a job I couldn't have done for love or money, placed a rope around the neck of our lovely lifeless horse, tied it to the end of his horse-drawn float, and dragged the body out of the stable. Then, letting down the back of the float, he drew out some chains and tied them around the stallion's neck. Next he mounted the float, and started to wind a handle, drawing in the dead body of the stallion. Within a few minutes, the O'Keeffe's man had gone, leaving the blood-stained yard and stable.

My Dad went back into the stable. Leaning against the manger with head bowed, he started to cry. I went over to him, put my arms around him, and started to cry as well. He put his strong toil-worn hand down gently on my head, and pulled me in close to him. I was seven years old.

CHARACTERS

When we describe somebody as "a real character," we mean that the person is either entertaining in a show-off sort of way, or is not the full shilling. There were plenty of characters around when I was growing up. Some were well-known like "Bang-Bang" and "Damn-the-Weather," but there were other characters who would have been known only by the locals. We had our fair share of them coming into our house and yard. They told stories, played cards, drank endless cups of tea, and discussed and argued over nearly every subject under the sun. But when the arguments were over, the slates were wiped clean, and there were no hard feelings.

Things are different these days. We live in an age of conformity, compared to when I was growing up, strange as that might seem. To step out of line these days, whether by expressing a controversial viewpoint, or doing something silly at the Christmas Party, you get tagged and pigeon-holed, in a way that can affect your promotion prospects or even your job. That, I suppose, is part of the stress that people have to live with these days. In my younger days, people were "easier" about a lot of things they said and heard. But as jobs were so scarce back then, the chances of having promotion prospects to damage were slim. Imagine worrying about your tax problems, when you have no income …

Everybody knew Bang-Bang, a little wispy man who ran from one side of the street to the other holding a key, pointing it at people and shouting, "Bang bang." He brought a bit of fun to the street, jumping in and out of doorways, out from behind cars, shooting at the hordes of kids who followed him around shouting back "Bang bang," while the big people stood in their doorways or stopped on the street to enjoy the scene.

Then there was Damn-the-Weather. From what I remember, he looked normal enough. He would stroll along the pavement in the middle of a group of people who never knew or heard of him, and suddenly cry out, "Damn the weather," and bang his arms across his chest and under his armpits. Such sudden behaviour would frighten the life out of the strangers around him.

If these characters, who brought a lot of fun to the street, were around today, they would be whipped up quickly by the white-coated brigade. But we accepted what they did and laughed more with them than at them.

Mickey Joe, from the nearby dairy yard, was another character, but in a very different way from Bang-Bang and Damn-the-Weather. Mickey Joe was probably middle-aged when I knew him first, with a mouthful of gums except for one large rotten tooth sticking out at the front. Summer or winter, sunshine or snow, he was always dressed the same way, in a striped collarless shirt, an old open waistcoat, brown trousers with one black and one brown legging, and a cap perched on his head. Mickey Joe lived in the yard with the cows. His clothes seemed to get dirtier and dirtier with the passing years, and if you were ever in Sheehan's shop, after Mickey Joe had been there, you would have "felt his presence" by the smell.

Apart from Mickey Joe's attire and his "strong presence," the fact that his nose was constantly on the drip, and wiped continually by a forearm which glistened from the performance, would almost certainly ensure that he would be left sitting during a "ladies' choice." He milked the cows by hand and then distributed the milk by horse and float to the various shops and dairy parlours around. Bearing in mind that it was not pasteurised or treated in any way, we made a point of knowing where Mickey Joe sold the milk, and staying well away from there. Once he brought us around a bucketful saying, "Get that down you, it's from the new cows we just got in."

The Ma accepted the gift graciously, but poured it into the pig trough, not wanting to take any chances. The next day Mickey Joe brought around more of the new milk to ask Ma to churn up some butter, which she did, though she declined his generous offer to keep half the butter for herself. Afterwards, our butter churn was given a washing like it had never had before. But Mickey Joe, despite his short-comings, was a decent poor devil who would do anything to oblige.

Johnny Forty-Coats was another character who came our way, but, to be honest, I cannot remember him being controversial in any way, except for the load of clothes he wore summer and winter. I remember him walking along the street with a sack slung over his back, and a few kids running after him shouting out: "Johnny Forty-Coats," as if the poor devil didn't know his name, or the number of garments he had on.

Joe Mullarkey was a man who begged my Dad to rent him space in our yard, so that he could make icecream, and store the hand-cart. My mother said that he was a dirty looking oul' fella, and that my Dad should not let him inside the gate. But my Dad, making excuses for "the poor oul' man who just wanted to make a few bob to keep body and soul together," ignored her pleas, and partitioned off a big shed to give him a cheap place to rent. If there had ever been a competition for the snottiest, drippiest nose, Joe Mullarkey and Mickey Joe would have battled out the finish nose-to-nose, with victory going by the shortest of slippery noses to Joe. Unlike Mickey Joe, Joe Mullarkey wore a suit and a collar and tie. But like Mickey Joe's clothes, his too showed no signs of ever having been cleaned.

Joe made his icecream in the yard, just outside his shed, beside the dung heap. I don't know exactly how he made it, nor had I any desire to find out, particularly after the first batch came on stream. But I have memories of Joe bent over a big metal container, stirring and stirring for ages, as his nose dripped into the mixture, and midget flies from the nearby dung heap dropped into the container and spread out, looking like black powder, before being swept into the white creamy liquid. I was there on my own with Joe when the first batch was ready.

"There you are, all for yourself and free," said Joe as he handed me his first masterpiece made on our premises, another first for the Bolands. Then he added, "Here's another one for your Daddy."

"Can I have another large one?" I asked.

"Sure you can, you must want one to give your Mammy a treat as well," said he, as he packed as much as he could of the deadly mixture onto another cone.

I went into the house and gave the biggest one to Wally, and the other two to the Ma and Da, winking at the Ma and shaking my head at her, unnoticed by the others.

"That's the best icecream I ever tasted," announced Wally.

"Me too," added the Da.

The Ma, looking suspiciously at the icecream, and then back at me, had not touched it. The Da finished his, and saw the Ma's cone lying on a saucer.

"I'll have yours, if you don't want it."

The Ma passed it over to him. I asked Wally if he would like another one, and he said he would.

Joe Mullarkey, thrilled that his new landlord's family was enjoying his concoction, heaped more of it onto a cone, which I brought in and gave to Wally. Wally and the Da finished their second helpings and sat on either side of the fireplace, licking their fingers, and saying how delicious it was.

"Do you like Mr Mullarkey's icecream?" the Da asked me.

"I won't be eating any of his icecream," said I. The Ma stopped ironing and stared at me.

"Why's that, is it too creamy for you?" asked the Da.

"No, it's because his snotty nose drips into it, and the flies from the dung heap get mixed in it when he's stirring it," I said.

"Jasus, I'm after eating two cornets of the stuff and I'm beginning to feel sick," said the Da. With that, Wally started to retch. The Ma quietly edged me towards the door and slid the ironing table into the centre of the room, separating me from the other two.

Wally, uncharacteristically, was the first to get the message.

"You little brat," he roared, leaping to his feet. As I took off through the door, I heard him and the Da crashing into the ironing table as they lunged after me. By the time they reached the yard, I had released Rover, who was growling softly, ready for action. Nobody dared lay a finger on me.

The Ma, following them out, laughed: "That's my boy."

When the icecream was ready, Joe transferred the mixture in a tin to his brightly coloured hand-cart. He then put on a spotlessly clean white coat, which his sister, who worked in a laundry, did up for him. Then he went out onto the street looking like a million dollars, ringing a hand-bell and shouting, "Delicious icecream, stop me and buy one."

And the poor eejits stopped him in their droves and bought

his snotty fly-ridden icecream. Ever after that, when my Ma saw Joe setting forth in his spotlessly clean white coat, pushing his lethal cart full of germs, she would say, "All that glitters is not gold."

And whenever I hear that phrase now, I think of Joe and my mother, always with apologies to my mother, for I know that even now, she wouldn't want to have her name linked with Joe's.

Some of my friends used to say, "Aren't you the lucky one! You can get all the free icecream, toffee, and cakes that you want."

But there was no way that Mullarkey's icecream, even if he blew his nose elsewhere and managed to keep the flies out of it, would ever appeal to me. But it was fun, now and again to follow Joe on his travels, just to see who was eating the mixture.

Although Joe gave us no trouble in the yard, and paid his rent promptly, there was concern about his production methods, and if anyone got sick, we might be blamed in some way for it. Eventually, after a lot of pressure from the Ma, the Da asked him to leave.

But Joe did not depart quietly.

He "effed" my Dad from a height. This sent Mrs Horan, who had heard the outburst, scurrying for her rosary beads. Worst of all, Joe then threw a handful of the freshly-made personally-flavoured icecream at the Da.

The Da retreated into the harness room saying, "Jasus, I'll get typhoid or something from this."

Then the Da, with his long driving whip, the one with the silver ferrule and his name engraved on it, came back out to do battle with dirty Joe. Joe crouched behind his hand-cart, which was loaded with the deadly ammunition. The Da advanced slowly, swishing his beloved whip, snarling through gritted teeth: "Get out of this yard, you snotty-nosed oul' git." While this squaring up was taking place, the Ma and me observed it from the safety of the harness room. We thought it was hilarious.

But Joe was nothing, if not a gutsy fighter, and was not a bit impressed by the Da's personalised whip. Joe proved to be a crack-shot, with the icecream, for once more he forced the Da to retreat to the harness room, covering him from head to toe with the deadly mixture. The Da turned on me and the Ma, who by now were helpless with laughter.

"It's all right for you two," said the Da, "even the soldiers in the first world war had only cold steel, mud-filled rat-infested trenches, and mustard gas to put up with, but I've got to deal with this snotty-nosed little hoor and his disease-ridden slop." Then Da added, "How, in the name of Jasus am I going to get the little hoor out of here, without catching some incurable ailment?"

Before we could think of an answer, Joe unleashed another salvo which came straight in the window and landed *splat* on the Da's beloved patent leather harness. The Da rushed to close the window, and got another smack of the stuff in the face. We could hear Joe shouting from the yard:

"Come on out with your whip. You and ten like you wouldn't be able to put me out."

"That's it now," said the Da, as he started to tie old sacks around himself.

"What are you going to do?" asked the Ma.

"I'm going to grab that little hoor by the neck, and drown him in his own snotty icecream, that'll be a death he'll remember for a long time to come," said the Da.

The Ma got between him and door, afraid to let him out in such a rage. For all his faults, Joe didn't deserve to be drowned — or poisoned.

"I'll get him out," said I, squeezing out through the door and into the battle zone.

With that, Joe taunted my Da: "Are you sending the child now to put me out?"

Joe liked me. He patted my head as I went past him to the stables, while he continued to shout abuse at my Dad.

But when I came back down the yard with a viciously snarling Rover struggling against the leash in his efforts to get at Joe in response to my "Seize him," the icecream man, terror-stricken, bolted for the gate. Rover and I raced out after him, only breaking off the chase when we reached Ardee Street. When we got back to the yard, the Da was delighted and, forgetting Rover's dislike of him, bent down to give the dog a grateful hug. However, Rover responded with a vicious snap, which sent the Da running back to the safety of the harness room. Later, the Da filled Joe's handcart with his bits and pieces and pushed it down to a house in Malpas Street, where Joe had a room. Rover and me went with him, just in case Joe made an appearance.

Joe never sold his icecream up our street again, leaving behind a legion of disappointed supporters, who swore they had never tasted icecream like it before in their lives. I have no doubt they were right.

Another character around our way was a lady known as Big Bridie, a widow who lived on the top floor of a tenement house in Chamber Street. Bridie sat in a rocking chair in front of a big fire which blazed summer and winter, leaving her with big red blotches on her fat legs. She told people that she had one son working in England, but later we discovered that he was living as a guest of the realm, making mail bags, and that he would be doing so for a long time. But whatever about her son's working conditions, Bridie made a fortune. She had two big rooms broken into one, with a brass double bed in the corner. She sold clothes of all shapes and sizes, for big and little people, which she draped over ropes which were nailed to the walls. She also sold shoes, potatoes, other long-lasting vegetables, the odd bag of cinders, holy statues and pictures, as well as undertaking to procure nearly anything you wanted. In addition, she lent money. All of this business was carried out from her one room.

But for all her enterprise, Bridie wasn't a money grabber. The people she lent money to were mostly friends or customers who really needed it, and she was very understanding when repayment was delayed. She was unlike the professional money-lenders in this respect, who went around banging on the doors for their money, while the poverty-stricken people who couldn't raise the instalment hid inside, pretending to be out.

Some people said that Bridie slept in her rocking chair, and, judging by the smell in her room, I would say that she did. Also, I don't think she ventured too often towards the wash jug and basin which lay on the table by her bed. Bridie wore a big money bag tied around her waist. It always seemed to be full of money. Anyone who liked could walk through the open tenement door, and on into her room, which was permanently unlocked. One day Bridie's son came home. In no time at all,

Bridie sold up everything, except the rocking chair, and went off back to England with him. We heard nothing about her for a few years, until the son returned one day, and told us that within a week of leaving Ireland, Bridie had tripped and fallen down some stone steps, and had died within a few hours.

Big Bridie, Mickey Joe and Joe Mullarkey are only a few of the many characters we knew. They all contributed memorable moments, in their own way, and gave us a reason to laugh, which was so badly needed in those difficult times.

WRITING A BOOK

When I was about twelve years old or thereabouts, I wrote a book, or should I say hundreds of pages of scribble. A man who lived near to us worked in some company that sold paper, and after he gave us a big bundle of plain loose-sheet paper for nothing, I felt I had to write or draw something on it.

After first filling sheet after sheet with drawings of horses and dogs, I then started to write. At about that time I had finished reading a couple of adventure books about China. One of them, if I remember correctly, was called *Escape to Chungking*. Books and films about China and the Chinese had always fascinated me. I don't know why — maybe out of fear, or my Dad's prophesy that "The yalla race will rule the world." Anyway, with stubs of old pencils I started to write an adventure story about China.

The more I got into the writing, the more gripping and exciting the story became. If a prize was given at that time for pure imagination, I might have surprised the odd Christian Brother or two, and believe me, there were quite a few "odd" ones around. Morning, noon, and night, in between feeding pigs and gathering slop from door to door, I wrote my adventure story about one-eyed, knife-throwing, gun-shooting, throat-cutting evil Chinese pirates, as well as the good "whiteys," played by my old pal Errol Flynn and his girlfriend Olivia de Havilland. On and on I wrote: late into the night by candlelight when I was forced to switch off the light; first thing in the morning when dawn broke to give even a glimmer of light, and I dragged my weary body with the excited mind back to my labour of love. Eventually the book got finished in the most glorious manner, with Errol and Olivia sailing off into the sunset, waving and smiling to all the now happy villagers they had rescued from the cruel bandits who had persecuted and tortured them for years.

I then set out to read it to all our neighbours and friends, and indeed to anyone who would listen to me, thereby inflicting on them more torture than those Chinese bandits could ever have done. Sitting ducks for this torture were the tenants of my Dad's house, who felt that I should be humoured, just in case it might help if they ever got into arrears with the rent. Then the Ma and Da had to sit through it, though the Ma seemed to enjoy it and offered encouragement. It was a serious adventure book, but after the Ma broke her heart laughing through the first few chapters, it was necessary for me to first tell all future listeners of the story that it was a serious one and they should not laugh.

The laughing stopped all right, but then all kinds of facial expressions replaced it, together with a lot of coughing, throat clearing, spluttering of one kind or another, along with much blowing of the nose, and a lot of use of hankies. I was feeling great about the book, really proud of my achievement, so proud that I decided to give all previous listeners a second dose of the story. With the reading went a floor show, which consisted of a different voice for each character, accompanied by sword fencing, shooting of bandits and dropping dead onto the floor where necessary. I am sure there was many a listener witnessing my dropping dead act who wished it was the real thing.

But my confidence took a bashing from a big countryman, who farmed somewhere out in County Dublin. He called into our yard on his way to market with a load of turnips, to see if we had any manure which he could collect on his way back home. When I started to read to him, he told me to shag off. I then told him where to stick his turnips, and he in turn aimed a belt at me. The Ma then put him out of the yard. As he departed, effing and blinding out of him for all he was worth, I followed him down the yard with a few more suggestions as to where he should stick the rest of his farm produce, starting with his cabbages and working through to his sheep. But after he left, the Ma gave me a few clatters for being so cheeky.

Just when everyone had got fed up with my readings and accompanying actions, Mr Mullery came up with the idea that I should get the book published. He gave me an address to go to, somewhere on the Quays, down near O'Connell Street. I climbed the stairs to an office at the top of a building, knocked at the door, and entered when called in. A kindly looking girl with glasses and long hair sat facing me at a desk.

"Do you print books, Mrs?" I asked, in my best Cork Street accent.

"Let's see what you've got, pet," said she sounding more like a nurse than a publisher.

Quick as a flash I opened the twine from around the brown paper parcel, and produced my beloved loose-leaf book. I went straight into the reading of it, complete with the character voices and all the actions. After only a few minutes of the floor show, your wan burst out laughing, at which point I very nicely told her it was not a laughing story.

"Sorry," said she, reaching for a hankie, with which she blew her nose and wiped her eyes. I continued to read and act the parts, but after a few minutes, she interrupted me to say that she wanted someone else to hear the story. She left the office for a few minutes and returned with two men, all three of them grinning like Cheshire cats. They sat down, all smiles, while I went back to the start of the book to read.

After a few minutes there was more laughter, this time from the three of them.

"It's not a laughing story," said your wan in a high-pitched voice about to burst out with laughter, as she again blew her nose and wiped her eyes.

"Sorry," chorused the two men.

I continued to read and act. As I read on, getting more and more excited as I came to the parts I liked best, I noticed that all three had their hankies out, wiping eyes, blowing noses, and stifling laughter as best they could. This went on until I was about a quarter way through the book. I had reached the part where I climbed onto a chair, and with hand upraised, supposed to be brandishing a sword, I jumped off onto the floor, leading the charge from the "chap's" boat to the pirate boat. The sight of me jumping off their office chair must have been too much for them, for suddenly, like a well trained choir, they all three burst into hysterics of laughter, with your wan doubling up on the chair, almost touching the floor with her head while her body shook. The other two were almost as bad, one of them standing up and leaning face first against the wall, convulsed with laughter. The other fellow just laughed out loud and wiped his eyes.

The sight of all this joviality, at a most serious and exciting part of the story, was just too much for me. Finally realising that they were taking the mickey out of me, I told them red-faced with anger and embarrassment, to "Shag off," as they started off another round of laughter. They were still laughing for all they were worth as I grabbed my sheets, brown wrapping paper and twine, and ran out to the landing, calling them a crowd of shaggers. They followed me out, still in hysterics, as I headed for the stairs, red-faced and calling them names.

Suddenly all the papers dropped from my arms and went tumbling, gliding, down the four flights of stairs. Again more laughter from the three as I set about gathering up my precious story, while still shouting abuse at them. Eventually I arrived at the end of the stairs with all the papers bundled up and jammed any old way into the brown paper wrapping, as the eejits on the top floor looked down at me, still laughing their heads off.

When I arrived home and told Mr Mullery what had happened, he too started to laugh, but quickly collected himself to say that they mustn't be much of publishers to laugh at such an exciting story as mine. The story was never the same after that, one of the reasons being that I had not numbered the pages, and I just couldn't put them back together again in the right order.

A Love Story

During the 'Fifties I was a teenager, although the term "teenager" was not part of everyday parlance. I started work, and with my new-found independence, started going to dances. Although the fast dances like the quickstep and the foxtrot were fun, as well as the tango, what really appealed to me was the soft, romantic music of the slow waltzes. Although, as the fella said, I couldn't put a foot under me, nevertheless I loved going to dances.

I went mostly to the Crystal Ballroom in South Anne Street. I loved the atmosphere, I loved the music, and I loved watching the good ballroom dancers who were usually the first up on the floor. And although the fast dances were fun, the most popular were the slow waltzes, where people in love, people who thought they were in love, people who hoped that one day they'd be in love, drifted around in each other's arms, some with closed eyes, as Louis Mullen or Don Geraghty sang *Make Believe, Always, When I Fall in Love, Secret Love, Smoke Gets in Your Eyes, Cara Mia, What'll I Do, The Tennessee Waltz* or *Only You*.

It was in the Crystal that I met a beautiful young girl on her eighteenth birthday, and went with her on my very first date to see *The Glenn Miller Story*. And although we shared a lot of precious and loving moments, it was not to be that we should journey through life together.

I was a hopeless romantic then, as were, I would say, most young people. I am still a misty-eyed romantic, despite life's slings and arrows. Being a Cancerian, I look back quite a lot, and a piece of music from the past can whisk me back in memory to almost taste, smell, and drink the atmosphere of the time. And often, recalling a special moment, or thoughts of a special person now gone out of my life forever, I am afraid that the tears will come. In these ungentle times, when it is considered

172

a sign of weakness to say "I'm sorry" when you're wrong, or for a man to shed a tear over a special memory, I am proud to admit that by this soulless standard, I am often weak.

I believe that music has a profound influence on people's moods and about the time when The Beatles arrived on the scene, the soft romantic music began to disappear from the dance halls and the hit parades. The hard music then took over. People no longer held each other gently and softly, at dances, or held hands when they went out walking. The hard music became harder and louder, to the point that songs no longer told love stories, or if they did, the words were inaudible. The soft romantic times of the 'Fifties and earlier had helped to produce gentle caring people, people you could trust and be safe with. There are not many of this type around today, so maybe we are a species whose time has come and gone. But I must admit that there are some caring young people around in these difficult and dangerous times, with a concern for others less well off, particularly in distant barren lands.

When I was about eleven years old, I met Mary and Michael, and came to witness a real love story for the first time. As with many magical moments, or special encounters in life, I did not appreciate it for what it was at the time. It took another twenty-one years for the story to unfold, and more time after that to realise what I had witnessed.

My father did business with a man whose home was near the foot of the Dublin mountains. Every four weeks or so, my Dad would drive there in his horse and trap, and in the summertime he would often take me with him. While my Dad and Mr Byrne did their business, and later sat talking and smoking their pipes in those lazy summer days, I went off to play with the children of the area. There were about eight or ten houses, all detached with large gardens, a kind of hamlet, in which the children lived. Next door to Mr Byrne's house was the Donnelly house, with the biggest garden of all, almost the size of a football field. Mr Donnelly was a builder, and I was told that he had built all the houses. Both the Byrnes and the Donnellys had one child each. The Byrnes had a boy named Michael, and the Donnellys a girl named Mary.

At the end of the Donnelly garden there was an embankment about three feet high, from the top of which, going away from the house, was a gently upward-sloping hill. The children played mostly on that hill, and beyond, in a valley. As you climbed onto the embankment and started to walk upwards on a summer's day, you could see the top of the hill set against the background of a blue sky. As you got nearer to the top of the hill, the greenish-purple mountains beyond gradually appeared, replacing the blue sky as part of the horizon. When you reached the top, the experience of just standing there and looking long and quietly all around you, especially for an eleven-year-old from the city centre, was absolutely magical. With the exception of Mary and Michael who just loved the place, I don't think that the other children really appreciated it at all.

From the top of the hill, you could see valleys, a lake, an old ruined castle, a broken disused mill by a dried-up millpond, and the sea in the distance beyond. The children were friendly and we all played happily together. Being from the city, I was a bit special, or maybe just different, and this seemed to ensure that I was always given a warm welcome when I arrived. Certainly I loved going there and always looked forward to the visits, usually knowing a few days in advance that I would be going, and excited at the prospect. Mr Donnelly had bought the hill and a lot of surrounding land from a man whose name in Irish sounded like "Minden," and so the hill at the back of the house became known as Minden Hill.

From the first day I met Mary and Michael, I knew that they were in love, although at the time I didn't really know what being in love meant. We had all been playing around the old castle on the other side of the hill, and then decided to race back up the hill, over the top, and down into Donnellys' garden. When we finished the race, I noticed that Mary and Michael were not with us.

When I asked one of the kids where they were, why had they not raced, I was told, "Ah they're up there holding hands. They're in love you know."

I walked back up the hill and over the other side on my own to take a look at these strange people, for the only other people I had known to be in love, were people in the pictures and my half-sister Ellen and her husband. They were holding hands when I caught up on them, pointing out towards the sea and laughing about something.

"I heard you're in love," said I. They just smiled. "What's it like?" I asked.

They were obviously used to being teased about their condition, for my request didn't take a feather out of them. Mary looked at Michael, and he answered me.

"It's nice."

That was the only time I asked them about their love. Now, many years later, I feel that if you were to ask the same question of people truly in love, the most perfect and believable answer would be simply, "It's nice."

Over the next few years I was a regular visitor to Minden Hill with my Dad, each time receiving the same warm reception. I still had the same joyful anticipation about going, and got the same thrill when I climbed Minden Hill and saw the valleys and lakes and the old broken mill. As time went by, the love between Mary and Michael grew stronger. They knew that they would one day marry and live in a house which her Dad would build for them on Minden Hill.

My Dad's business with Mr Byrne wound down gradually, and finally finished altogether. But the Da and Mr Byrne remained friends, and occasionally we would still drive to the foot of the mountains and drop in, just to keep in touch. Each time the children, now teenagers, welcomed me warmly. But things were changing. As we grew up, the slots in the social scale were beckoning, for the children of the hill were in secondary school, with all the options that such an education provides, while I was still going from door to door for my father, collecting pig-feed. One of the last times my Dad and I went to the hill together was in response to a hand-delivered letter from Mr Byrne telling us that Mary had died.

I remember going to the funeral with my father in a black car provided by Tommy Rice from Maryland. It was in a small country church. After the Service, Mr Donnelly, Mr Byrne and Michael helped to carry the coffin down the gravel path from the door of the church, out through the gate, and on to the lane where the hearse was waiting to take Mary on her last journey.

> "There'll be no more walking, my love,
> Nor excited talking, my love,

> Of the home we'll build right
> Here on Minden Hill
> For the Lord above
> Called you away, my love,
> And once again time stands still
> Here on Minden Hill."

The graveyard was a short distance from the church and we all walked there behind the hearse. I have been to many funerals in my life, and of course as you get older you find yourself going to more and more of them, paying final respects to people you knew and loved, people you respected, and sometimes people you didn't know at all, but for some worthwhile reason you feel you should be there. What I remember about that funeral, through the hazy memory of time, is that there was great quietness at the graveside. The ceremony was dignified. Only Mrs Donnelly seemed to sob, briefly, and very quietly, like a whisper in the distance. Opposite me on the far side of the grave, stood Michael, tearless. On my right, at the top of the grave, stood a gentleman reading from a book as Mary's coffin was lowered into the earth.

> "The Lord is my Shepherd, I shall not be in want,
> He makes me lie down in green pastures."

And as the clay was dropped onto the coffin with a finality of sound unequalled in my experience, the gentleman read on:

> "Surely goodness and love will follow me
> all the days of my life."

This last bit I couldn't understand.

> "And I will dwell in the house of the Lord forever."

A few months later my Dad and I went out again to the hill to discover that the Donnellys had moved away, and that Michael had not spoken nor moved out of his room since Mary's death, except to visit the hill each day. We also learned that Mary had died from leukaemia, a disease that hit her suddenly and overpowered her with devastating speed. On the day that herself and Michael were due to start University, she lay dying in hospital. On Mary's death, Michael practically gave up on life.

For about two years after that, I called out a few times to see Michael, only managing to see him on one occasion.

"How are you?" I asked, putting out my hand to him.

He didn't answer as he took it and looked sadly at me. Then he shook his head and walked away. That was the last time I ever saw Michael. He was unable to cope without his beloved Mary.

I was living in England when Michael died at the age of about thirty-two. He died in the same month, and on the same date of the month as did Mary, and it might have been at the same time, for they both died in the early hours of the morning. I learned this about two or three years later, when home on holiday I bumped into one of the hill friends in Bewleys. He told me that Michael neither cried nor smiled since the day Mary died, but when found dead in bed, he looked happy with his arms extended, as if he had been reaching out to someone he was pleased to see. He also told me that since Mary's funeral, Michael had every day, come hail or shine, walked up to the top of Minden Hill, stood there for a little while looking around him, and then walked down again.

I think of Mary and Michael from time to time, when as happy children we first met and stood on the top of Minden Hill, looking towards the lake and the sea beyond, dreaming our dreams, as the perfumed breezes freshened our young, shining, hopeful faces. Some years later, on my return to live in Ireland, I tried to find Minden Hill, but in the general area of where I thought it was, all I could find were rows of new houses.

THE GHOST FROM BROWN STREET

With the exception of going to the pictures now and then, our entertainment in the winter evenings was mostly home-made, telling stories, playing cards, and having the odd sing-song. The story-telling covered all and every subject, but ghost stories, and to a lesser extent fairy stories, predominated. There was a story that a man with a cloven hoof appeared regularly in a pub in Marlborough Street, and if you let him buy you a drink, you would disappear from the pub and end up in hell. If it was one of those non-alcoholic beers, who would want to be seen in the pub again anyway?

Then there were stories about devils playing cards in the Hellfire Club, up on the Dublin mountains. It was said that you should bless yourself three times and say three Hail Marys when passing a graveyard, or a few of the residents might jump out on you. Throw in the odd Banshee or two, a headless horseman riding through Crumlin Village at midnight, and umpteen other such stories, and you had a solid foundation for fear of anything that moved or rustled when you were out in the dark.

Tommy Rice ran a taxi limousine business from his yard by the canal in Maryland. From time to time, Da would hire a car for special occasions like weddings or funerals. One evening the Da came home and told us that one of Tommy's two drivers had died suddenly. He described the man, whom I readily recognised as Mr Kelly. About a week later, just as it was getting dark, I was walking down Donnelly's Lane on the way to Brown Street, taking a message to someone for my Dad. Just as I reached three-quarter ways down, suddenly the "dead" driver turned into the lane, walking straight towards me. I nearly fainted, my heart began to pound, my legs went weak, and the perspiration started to roll off me. I was terrified. As he came nearer, his head bowed deeply in thought, his footsteps made not a sound, or, if they did, the pounding of my heart drowned out the noise. He moved on silently towards me and then past me, without making a sound, while I leaned against the wall paralysed with fear. When, whimpering and shaking, I managed to look behind me and after him, he was nowhere to be seen, he had disappeared into thin air. When I recovered, I ran, or rather staggered, back home in the other direction, by way of Brown Street, Weavers' Square, and Ormond Street. I ran into the yard, and on through the harness room into the living-room where I collapsed in front of the Ma and Da, who were sitting by the fire.

"I've seen Mr Rice's dead driver, I've seen his ghost, he was in Donnelly's Lane and he walked right past me," I shouted. I was in a terrible state and the Ma and Da, obviously concerned, hugged me and did their best to calm me down, while I was shaking and sobbing. I recovered after an hour or so, with a lot of loving tender care, and despite my terrible experience, managed to get to sleep under the blankets, while the Ma slept in an armchair beside my bed, just in case the ghost showed up again.

The next day I told everyone about my experience and I became something of a celebrity, with friends and neighbours anxious to know all about it, and telling me how brave I was not to have run away but to have stood my ground and let the ghost walk past me.

Ghost stories were such a part of our lives, that no one, including the Ma and Da, disbelieved my story. That evening after tea, at about the same time as I had seen the ghost the day before, I went into our yard. I could hear my Dad's voice outside and being curious, I meandered down the yard and out through the gate to see Da talking to a man who had his back to me.

"Here he is now," said the Da. "Tell this man about seeing Mr Kelly's ghost in Donnelly's Lane last night." When the man turned around to face me, I nearly died, for it was Mr Kelly, my ghost.

"It's him, it's him, it's Mr Kelly's ghost," I screamed as I collapsed against the gate.

"That's not Mr Kelly, it's Mr Doyle, and he's got a few good years left in him yet," said the Da, putting his arms around me

Taking the Horses to the Forge

At one time we owned about six horses, including the black mare which was kept for Sunday driving and light yard duties. Some horse owners were very careless about their horses' feet, but not my father. Every horse had to be properly shod at all times, and this meant lots of visits to the blacksmith in the forge.

When I was about seven, or even a little younger, I was trusted to take the horses to the forge, usually leading them by a rein to the winkers. As far back as I can remember, all the horses we ever owned, were "true," as my Dad said. That meant they were good workers, and would not bite or kick. Some horses that have been mistreated can be dangerous, in that they might well kick out with their hind legs if you passed behind them. For a horse to kick out is more a protective than an offensive gesture, for horses are not generally dangerous animals. If a horse has a tendency to bite, that is a fairly good sign that it has been beaten about the head.

Once my father bought a horse which had been mistreated and nearly starved to death. I was with my Dad the day he bought it. We were driving our horse and cart down Queen Street, towards Bridgefoot Street, when the Da suddenly pulled up the cart, and climbed down to look at what could only be described as a bag of bones tied to a lamppost outside a pub. Somehow the poor unfortunate animal mustered the energy to kick out at the Da, and tried to bite him as he carried out an examination. My Dad then went into the pub, and came out almost immediately with the owner, a travelling man. Quite a lot of haggling went on about the price of the horse, during which the owner several times spat on his hand and held it out to my Da, the idea being that if the Da agreed the price, he would seal the deal by spitting on his hand and slapping the two hands together. Although I had often seen my Dad participating in this ritual, he did not do so on this occasion. When the price was eventually agreed, the Da just held out the money, with no hand-slapping.

In accordance with the custom of the time, the owner would give back a "luck penny" to the buyer, which could be a few shillings or a few pounds, depending on the price paid. On this occasion, when the seller extended his hand with some coins in it, to my surprise I saw the Da strike the hand away, sending the coins over the street. The Da then told the traveller that he shouldn't be allowed to own a horse because of his cruelty. After a heated exchange, the Da untied the horse and led it away. I drove our own horse home at its usual fast walking speed, but the Da led the new horse home very slowly, for it was tottering a bit, too weak to walk fast.

This was a horse my father did not want, but he couldn't bear to ignore it, to leave it with the traveller because of the way it had been treated. My Dad had many friends in the travelling community, fine people who looked after their horses, and who often called to the yard to smoke a pipe with him, and talk horses together. After about three months, the Queen Street horse was back to full health, with a good strong body and a lovely shiny coat. Good on you, Da.

Nearly every week, one or other of the horses would need its shoes tightened or replaced, or a new set of pads. It was my job to take it down to the forge, either after school or on Saturday mornings. Although the Da preferred me to lead the horses to the forge, I preferred to ride them. Some horses, freed of the weight of the harness and cart, and with the added unfamiliarity of a rider perched up on their backs, might become frisky going along the road, which could be dangerous to other road users, especially if they bucked or kicked. The real danger, however, was when some gang of kids, seeing me riding on my own, would try to frighten the horse, just for the *craic*. More by luck than anything else, I never had a serious mishap in all my years going up and down to the forge.

We went mainly to one blacksmith in Island Street, just off Bridgefoot Street at the Queen Street Bridge end. Sometimes when I got there, there would be a line of seven or eight horses stretching around his yard. The horses were tethered to iron rings set into the wall, leaving the owners free to lounge in the sun, chatting or playing cards, or to retreat to the warmth of the forge if the weather was bad. I always stayed inside the forge, because I loved to see the blacksmith at work. The "smithy," a thin wiry man, worked at a steady pace, concentrating on his work without much small talk, but he was a pleasant, likeable man.

All around the walls of the forge, there hung pieces of steel of various widths and lengths. Whereas all our horses, with the exception of the black mare, were heavy working types, requiring the same kind of shoes and pads all the time, the smithy had to keep stocks of iron on hand to fit all sizes of feet, for donkeys, ponies, mules and horses. Although his main business was shoeing horses, the smithy also made iron gates, trolleys, trays — in fact anything you could imagine. You had to watch where you walked in the forge, for fear of tripping over one of his half-finished contraptions.

At the end of the forge, furthest from the door was the fire, the centre-piece of the whole place. Just beside it stood the anvil. Hanging down from the ceiling was an iron bar with a ring at the end to which the smithy tied the horse's rein while he worked on its feet. The fireplace was a brick structure about three feet high, and about five or six feet square. Affixed at the side, about five feet off the ground near the top of the fire, was a bellows, its long wooden shaft protruding about five feet. The fuel was smith's coal, or slack, and the fire enclosure was always filled right up to the top.

Each horse's requirements were usually different. If a full set of shoes or pads was required, the smithy would take off the old ones and examine and clean down the hooves with a rough file. He would take a piece of iron down off the wall, cut it to the required length, then grab it with his specially long pair of tongs and thrust it deep down into the fire. Next he would pull quickly on the shaft of the bellows, which made a kind of humming noise as the wind from it reddened up the fire. After a few minutes of this, the smithy would plunge the tongs back into the centre of the fire and retrieve the now flaming red iron. The iron would then be transferred to the anvil, where it would be shaped to the approximate shape of a horseshoe, and stretched at one or two places, either at the front or side, to make a little clip to help hold the shoe in place. Next the shoe would be spiked by a tool used to press the roughly shaped burning hot shoe up against the horse's hoof, thereby moulding its shape to help the smithy adjust the shoe to fit. Then the shoe was returned to the fire and the process repeated until the smithy was satisfied that it fitted. After that the nail holes were made, and if the shoes needed any special refinements to help the horse balance or walk more comfortably, these were added.

When all the shoes were made, they were then nailed on, together with the pads, and finally the hooves were dressed down and oiled. The pad was a kind of rubber block, mounted on thick canvass-like material which was nailed on between the shoe and the hoof. The thickest part of the pad sat to the rear of the shoe, just under the horse's heel, to stop the iron shoes slipping on the road. The shoes were usually made a little bit smaller than the actual size of the horse's hoof, to allow the hoof to be dressed down and given a smart appearance. The tool used to remove the old shoes was called a buffers. It had a flat side to break off the top of the old nails, and a pointy side, or end, to hammer out old nails to the point where they could be drawn with a pincers or claw hammer.

The forge was always hot because of the fire, and though uncomfortably hot on a summer's day, it was cosily so on a bitterly cold day. Entering the forge on a snowy winter's day was just heaven. Added to the heat was the ever-present smell of the burning hooves, and the smith's slack, when dampened in a bucket of water and added to the fire, sent a twirling bluish haze of smoke spiralling to the roof, giving off a sweetish smell. The day after a visit to the smithy, I could still get those tingling smells which lingered in my coat.

On occasions, when the smithy was not too busy, he would let me do little jobs for him. I started by using the tongs to pick up the iron, putting it into the fire, and then heating it up by pulling on the bellows shaft, which I reached by standing on a butter box. As time went on, the smithy would let me take off the horses' shoes and strip down their hooves. I would examine the hooves carefully to see if any stones or dangerous objects might have become lodged in the centre, which I would remove with a hook-bladed, bone-handled knife. Next I would flatten off the hooves with a large thick double-sided file. The smithy taught me how to make perfectly good pads from old car tyres, then how to nail shoes on the proper way. Horseshoe nails have a slant at the tip, and, if not hammered in properly, they could go right up into the horse's foot, causing pain and crippling the horse until the injury healed. When hammered in the proper way, with the slant of the nail facing outwards, the nails emerge outside of the hoof, where the tops are twisted off with a claw hammer, and then rasped down to nick or grip the hoof and keep the shoe tightly in place.

In time, the smithy showed me how to cut the iron bars, fire and shape the shoes on the anvil. Then I was shown how to spike the hot shoe and try it against the hoof to get the size right.

Gradually I went through the whole process from beginning to end — making and fitting a set of shoes for one of our own horses. I was no more than twelve at the time, and it was a great thrill for me to lead one of our horses out of the forge with a set of shoes and pads which I had made from beginning to end. The smithy was very proud of me too, but then I *had* served a six- or seven-year apprenticeship to the trade, on and off. The next time I called in, the smithy had a special present for me: my own made-to-size leather apron.

The easiest part of the shoeing operation was removing and dressing or tightening up shoes. Loose shoes were tightened by removing broken nails, which had caused the loosening in the first place, and replacing them. The tools used to do this were the buffers, pincers, file and claw hammer. Looking back, I can see that I had a liking for the work, which it would have made sense to pursue, given my background. I never mentioned the smithy's training at home, and it was only by accident, when I was about twelve, that the Da found out about it. In a moment of weakness, I displayed some of the skills I had learned, and lived to regret it for the next few years.

Winter brought its own special problems for my Dad and his horses. It was not just the freezing rain which drenched the Da and the others who had to drive in it, but also the frost, which made the roads treacherous for both horses and men, and made the shovels and pickaxes painful and difficult to hold. But whatever about the many miseries that winter brought, the one which affected man and beast most, as far as we were concerned, was the frosty roads.

My father listened tirelessly to the weather forecasts on the radio each night before going to bed, and then stood outside, assessing the weather for himself, before making a decision as to whether or not frost nails would be needed in the horses' shoes.

Frost nails look like ordinary horseshoe nails, except that where the normal horseshoe nail is flat-headed and counter-sunk into the shoe to provide a continuous level surface, the frost nail has a raised triangular shaped head — just like "Killer" Kelly — which protrudes about a quarter inch from the shoe. These nails grip the road surfaces, through the frost, ice, or snow, and enable the horse to walk and pull heavy loads without slipping.

In icy weather, some of the standard nails had to be knocked out and replaced with frost nails. Although only every second ordinary nail would be replaced, removing four old nails per shoe and replacing them with four frost nails, meant eight operations per foot, thirty-two per horse, and over one hundred if four horses were going out to work next day. And as the frost nails usually wore down flat in one day, that meant that the operation had to be performed every night of the week during the frosty season. On occasions when the horses went out without frost nails, and the temperature dropped below freezing, not only would the horses be unable to draw even light loads, but they would have difficulty in keeping their feet under them in order to get home safely. There was no real substitute for frost nails, but sometimes, just to get a horse home, the driver would tie old sacks around a horse's feet.

When frost nails were suddenly required, it would be almost impossible to get a blacksmith to put them in for you at short notice, such were the demands on his time. With petrol not readily available during the war years, businesses relied heavily on real horse power to get their products delivered. Many blacksmiths were contracted to firms, on stand-by for nailing duties when frost came. In the circumstances, a lot of small business people who needed to go out with horses to make a living had to put in their own frost nails, in the same way that farmers carry out running repairs to a whole range of farm machinery on a day-to-day basis. But some horse owners did not have basic blacksmith skills, and were stuck if the smithy couldn't get to them.

For some reason or other, my Dad, who was quite a handy man in his own right, could not put in frost nails, or so he told us. Instead, he called on Wally to do the job. Wally was very careful, to the point of being painfully slow. This meant that he would work half the night putting in the frost nails, and then go out to work the next day. If the Da made the decision early in the evening to put in the nails, then Wally could complete the job reasonably early, and get a good few hours' sleep.

One evening, watching Wally flootering around trying to get out some flat nails from a horse's shoe, I said to him, in that moment of weakness I told you about earlier, "Here, give me that buffers, you're breaking my heart looking at you."

I then took over and knocked out the flat nails, hammered in the replacements and filed them off in a fraction of the time he would have taken. While I was doing this, the Da was looking on in amazement. When the horse was finished, it was led out and another one was led in. In less than an hour I went through all the horses, turning out first-class jobs on every hoof.

When I was finished, the Da said, "I didn't know that you could do that, son. Where did you learn to do it?"

When I replied, "Sure I could do that when I was eight years old, and I can even make a set of shoes in the forge and put them on," he gave a big smile. At first I thought it was a smile of pride. Later I discovered that it must have been a smile of contentment, for he realised at that moment that his problems with regard to having to rely on only the "perfect" one to fit the frost nails were at an end.

Thereafter I dreaded cold frosty winter nights, when at about 4 am, in the middle of a deep sleep, maybe dreaming of the delectable Rita Hayworth, I would be woken by the Da's whisper:

"Wake up, son, I want you to put in a few frost nails for me."

Cold and bleary-eyed as I left my warm bed to go out into the chill of the stable, I could hear Wally snoring contentedly. But there was no way I would let the Da down, a man who loved me and went out to work for me, come hail, rain or snow, despite what he called me, or threw at me from time to time.

THE TRAVELLERS

A lot of people travelled in those days. There were the priests from Mount Argus who travelled around twice a year collecting dues, the travelling moneylenders, a man who travelled for a pawn office, the travelling dentist, and the travelling tin smiths who would do a beautiful job of repair on pots and kettles. There were also the travelling tradesmen who cycled around with their tools strapped on their bikes, ready and able to solder a pipe, glaze a window, make a door, fix furniture, build a wall, paint a house, or do any one of a thousand odd jobs.

But my favourite of all was the travelling musicman, who went from door to door selling the printed words of popular songs, and sometimes singing them himself. One particular musicman named Mikey had a lovely tenor voice, and people on the street would often ask him to sing a few of his songs.

Then some smart-alec might say, "Let's hear the first one again," and away he would go singing it once more. At the end of that they might say, "Ah, I don't like any of them."

When this happened you would hear Mikey in a different kind of voice altogether. But everyone liked him and welcomed him when he came around. He always got a warm welcome from my mother, because she liked to sing, and Mikey was a Northerner like herself.

I remember one particular time that Mikey called to our yard. A few neighbours had gathered and were chatting to the Ma and Da. This was one time when Mikey had a captive audience. In my mind's eye I can still see Mikey as he stood just outside the entrance to the hayshed, half-way between the dung heap and the swill trough. With head thrown back, battered cap perched to one side, eyes closed, thumbs clipped into his braces, he gave us all he had. The neighbours hushed and listened while his beautiful voice filled the air, and after each song he was greeted with a hearty round of applause.

When he had performed seven or eight of the songs he had for sale, he asked, "Will you sing them with me?" And ourselves and the neighbours joined in with Mikey and had a right old sing-song.

Of course, those who wanted to join in the sing-song either had to know the words or buy them from Mikey. And at a penny a song for seven or eight songs, he did well in our yard. But what was beyond price was the sight of my parents and friends standing happily singing Jimmy Kennedy's lovely songs, like *Red Sails in the Sunset*, *The Isle of Capri*, or *South of the Border*. Mikey brought us not just music, but also a sense of fun. And that sense of lightness and fun was a welcome respite from the poverty which existed for some, poverty which would be hard to imagine even in the worst conditions of today. The simple fun and good neighbourliness which prevailed at that time made even the worst poverty easier to bear.

To pay out seven or eight pence for Mikey's songs did not come easy to people, and not everyone could afford them all. But they got great enjoyment from the songs.

Long after Mikey had gone, the words and melodies of his songs wafted out of open windows, sung by singers and non-singers alike, by messenger boys on bicycles, and by gangs of people coming from the pictures late at night. Four or five abreast, arms around each other's shoulders, they would stroll along singing their favourite songs at the tops of their voices. And if the travelling chorus dallied a bit too long singing under a lamp post at a late hour, some nearby window would open and a voice would shout, "If you don't get the hell out of here, I'll throw a bucket of water over the lot of you."

Sometimes this brought a good-natured response from the chorus, which would move over to the window, changing the song to fit the occasion, like *Powder your Face with Sunshine*.

People who couldn't afford to buy all of Mikey's songs would learn the ones they had bought, and swop them with neighbours for the ones they had not been able to buy. When Mikey completed his performance and made his sales, my mother would bring him into the house for sandwiches and tea. And when he was leaving, she'd pack some food for his journey.

"You can't beat the Northerners for singing," my mother would say as she and the Da waved him goodbye.

The travelling dentist was something else. If the travelling musicman got the neighbours to sing, and left a nice musical feeling behind him, the dentist did the opposite. News of his arrival in the street was passed on bush-telegraph style, from house to house, warning anyone planning to avail of his services to wash out their mouths, sharpen their gums, or simply get ready for pain. One particular dentist, a dirty-looking oul' fella, with bad breath which would knock down a bull elephant at thirty paces, visited our street a few times. He called to our home just once that I can remember, invited by the Ma to do a job on me. And it was in the thirty seconds I spent standing in front of him that I encountered his breath.

I always think it is reassuring if people selling a particular service can demonstrate that they themselves are getting the benefits they promise you. The dentist's bad breath and rotten teeth reminded me of a bald-headed barber from Meath Street who sold hair restorer for four pence a bottle. For those with hair, he provided a haircut and a bottle of hair oil, all for four pence. The hair oil kept my hair in place and my bicycle oiled.

When we knew that the dentist was in a particular house, we would peer through the windows to see what was happening. Even though we were usually chased away, we would sneak back, fascinated, and put our ears to the keyhole to hear the screams of pain. I was about seven years old when the man himself made his one and only visit to our house.

For days I had had a loose tooth which caused me a lot of pain. Mr Mullery said it would fall out in my sleep one night, and if I put it under my pillow, the fairies would take it and leave me a penny in its place. That seemed to be a very civilised way to solve the problem, and I was prepared to give it a go. But that wasn't good enough for the Ma.

"Come into the parlour and let the gentleman see your loose tooth," she said.

"No, it hurts," said I, getting a whiff of your man's breath.

"I won't touch it, only look at it," lied the dentist.

I reluctantly agreed, opening my mouth wide, but very stupidly closing my eyes when asked. Like a flash, a smelly unwashed hand shot into my mouth, grabbed the tooth, and pulled it out. I let out a string of curses. The eloquence and vocabulary of my outburst made my poor mother go crimson. Your man blessed himself and covered his ears.

I fled howling out to Mr Mullery in the cake shop. I showed him my poor bleeding gums, and sobbed out the story of how the dentist had tricked me. Mr Mullery, always the gentleman, dried my tears, gave me a soothing drink, and generally quietened me down. Then, taking me by the hand, trailing the usual cloud of flour, he headed for our house, saying, "Come on, let's talk with this man."

When we arrived in the parlour, the dentist was sitting white-faced on the couch, while the Ma sat in front of him holding a glass of water. Mr Mullery, always my supporter, "savaged him verbally," as he was later to say. In his best Irish-American accent, he said, "How dare you, sir, hurt this small boy." I was surprised and disappointed at Mr Mullery's attack, for I expected him to give the dentist a dig in the eye, or at the very least call him names. Whatever "savage" signal the words conveyed, the dentist immediately upped and went without a word, and was never again seen in our house.

The three of us watched in silence as he left. Then my mother stirred herself and said, "I'll make a cup of tea."

"Good idea," said Mr Mullery, "I'll make a few pancakes," and off he went, me running behind him, safer with him than alone with the Ma who might remember my curses at the dentist, and give me a wallop.

A few minutes later, the three of us sat calmly as if nothing had happened, eating pancakes smothered with butter and sugar, and drinking mugs of tea. Other than my mother saying, "The pancakes are lovely, Mr Mullery," and he replying, "That's a grand cup of tea, Mrs Boland," nothing was said until we started the second round of pancakes.

Then my mother looked at me and said, "By God, you're some offspring. I'd drown you, only the gypsy whose hand I crossed with silver told me you'd be a great man one day, with your own bank account and briefcase." Then Mr Mullery started to laugh, and my Ma and me joined in. All was forgiven.

The travelling dentist pulled teeth for two shillings a tooth, but if you had more than one to come out, he would do a deal. At that time, nobody was too concerned about saving teeth, even if they only needed a little maintenance.

"Ah, take the whole bloody lot out, the hell out of that," he was told, and there was no better man to do the job at two shillings a go, less the quantity discount. Some people liked to hold on to their front teeth, but many front teeth, particularly in older people, turned brownish-green, and when this happened, the romantics among them could at best only hope for a hug from the opposite sex.

The travelling dentist, so far as I could see, performed just two functions, pulling teeth and fitting dentures. I was told that some people got all their teeth out at one sitting and, despite the bleeding and painful gums, got a set of dentures in there and then. You could choose the dentures from the sets the dentist had with him. Some people believed that if you got the teeth out and the dentures in at the same session, this would ensure a better fit than waiting for the gums to heal.

With dentures being chosen off the shelf in this manner, it is no wonder that many people could not wear them full-time, leaving them instead on top of wardrobes or in bedside jars. I often wondered where the ready-made, sometimes second-hand looking dentures might have come from, for it's hard to imagine that they could be mass-produced in the same way as shoes. Some said that the dentist had a deal with the morgue, and the undertakers for the collection of "left-over" dentures, which might account for the appearance of some of them.

Some people just wore one part of the set, either upper or lower. Others wore them only on Sundays or to weddings or funerals. On special occasions they were all right for drinking with. But when the food was served, the dentures went into the pocket. Hard foods, like meat, would be cut up small enough for "gumming," or else swallowed unchewed. Some people persevered with badly fitting dentures and became quite proficient at eating with them, although in some instances, what were once big open smiles became evil-looking grimaces due to the facial contortions necessary to hold the teeth in. And then there were those who did not have complete control over their dentures, allowing them to fall out at the most inappropriate times. Such a person was Tommy Hayden, whom the Da invited around for dinner, but mostly to introduce him to a nice girl from The Tenters.

Tommy was a very decent man who wanted desperately to get married, and this was the Da's one and only attempt at match-making. Tommy's problem was that he was shy, and too embarrassed to take off his cap because he was bald. That night at dinner, Tommy, still wearing the cap, grew more and more confident in his conversation with Deirdre, who seemed to be very interested in him. Tommy was expressing his views more strongly, more excitedly than he had ever expressed them before. Sensing a conquest at last, Tommy opened his mouth just that bit too wide, for out fell the top dentures, slap-bang onto the plate, into the middle of his roast chicken, potatoes, and peas. Deirdre had a sensitive and uncontrollable stomach, and when she saw a set of top dentures grinning up at her from between a potato and a chicken leg, her stomach reacted, with disastrous consequences. The Da, who sat next to her, got the worst of it.

We suspected that Tommy's loose dentures and continually capped head would be too much for *any* woman.

Happily, we were wrong about this, because a few years later Tommy married a very nice woman, also sporting a set of "delph," and a big hat which she seemed to wear day and night. We often wondered if they ever took off their hats in the house, or if they got excited from time to time over their roast chicken, roast potatoes, and peas. Deirdre married a fellow with a mighty head of hair, and a mouthful of teeth that looked like his own.

In the way that songs can bring back memories of people and situations, I often wondered if Deirdre could ever again face roast chicken, and if so, did it ever remind her of Tommy's top teeth.

MR CAREY

We did not have many heroes around our way, but Mr Carey
was certainly one of them. He drove a horse-drawn bread van
for Johnston Mooney and O'Brien, delivering bread and cakes
to shops in the area. His van was high-sided, completely
enclosed, and balanced on two large wooden wheels with
thick steel rims. The horse, which pulled what must have been
a very heavy weight when fully loaded, was an Irish draught
horse, 16 to 17 hands high, and always magnificently turned
out, with polished harness and brassware, and hooves gleaming
with oil.

When Mr Carey delivered to Sheehan's shop, he sometimes
pulled his horse and van in beside our gate, and carried the
bread and cakes across the road. The arrival of his van each
morning was an event for us kids. Mr Carey would go to the
back of the van and release the massive bolts, which allowed
the two big doors to swing open. Then he would take out trays
of loaves, turnovers, ducks and pans, and carry them across to
Sheehans, trustingly leaving the van doors open. Next he
would open the large drawer at the bottom where all the cakes
were stored. Our eyes would be popping and our mouths
watering, as he picked out the cakes which Miss Sheehan had
ordered. Later we would go into Sheehan's shop, just to gaze
at the same cakes through a glass cover. My favourite cake was
"gurcake," a dark brown slab full of currants, raisins and
sultanas.

What I remember most about Mr Carey's deliveries is the
moment when he would fling open the doors and that special
aroma of fresh bread came rushing to my nostrils. To this day,
because of Mr Carey's van, one of my favourite smells is that
of fresh bread, and when I come across one of the shop
bakeries, I cannot help walking in and around, and usually
buying a loaf. The darker the crust, the better I like it. An estate

194

agent told me once that the aroma of bread baking creates a homely atmosphere in a house and is a great aid to selling it.

During the second world war, which was referred to in Ireland as The Emergency, the bread which we got was practically black. North American wheat was unavailable, and we had to use native wheat instead. Because of our unpredictable climate, our farmers had great difficulty in producing good wheat. When they nearly had it dried and decided to give it another week to be sure, down would come the rain. The dampness of the wheat, plus the fact that only wholemeal flour was being produced, combined to make the bread both black and doughy. Some people had the foresight to lay in a good supply of good white flour prior to the Emergency, and others managed to acquire it somehow, either baking the bread themselves or getting a baker to bake it for them. After the war, building on our experience, we developed very efficient wheat-drying systems, which people came from all over the world to study.

It has been said that the black bran bread sold during the war years was the healthiest bread ever made in this country. Today, people buy bran separately in health food stores to supplement their diets, and pay a lot for it. I well remember seeing white bread for the first time. It would have been in the 'Forties. I was standing at our gate when a girl walked up the far side of the street with an uncovered white loaf under her arm. Some of the neighbours stopped her and gathered around to look at the loaf, which was unbelievably white, although probably no whiter than today's bread. The girl passed the bread around for each of us to inspect, and one man took a photograph of her holding it high above her head, the way a county captain holds up the Sam Maguire Cup after the All-Ireland final. We stood there handling this magnificent bread, most of us with dirty hands, and marvelled at its colour. Some of the people ran down to Meath Street to get some for themselves, only to find that it was all sold out. In a few days, though, we all had plenty of the white bread, and we ate it everywhere, even walking along the street, dipped in dripping or covered in jam.

Mr Carey had three claims to fame, in our view. One was that he was a nice man, who trusted us with the open doors of his bread van. The second was that he had almost become wealthy, to a degree beyond our wildest dreams, but, through a strange quirk of fate, had thrown away his chance.

Every year, for many years, Mr Carey had bought a book of tickets for the Calcutta Sweepstake. But one year, he decided not to buy them, and passed them on to another purchaser who won the magnificent prize of £10,000. But it was Mr Carey's third claim to fame which set him apart from all others, for he was the father of Johnny Carey, probably the greatest soccer player ever to come from this country, and one of the finest the world has ever seen. The late Sir Matt Busby, President of Manchester United Football Club, never one to throw out compliments lightly, said of Johnny, "He was one of the great players, and I would go so far as to say one of the best I have ever come across. He was a great ambassador for our club."

Over a five-year period during the late forties and early fifties, Manchester United came heartbreakingly close to winning the League Championship, once finishing fourth and four times coming second. Finally, captained by Johnny Carey, they won it in 1952, after an interval of 41 years. The serious minded *Guardian* newspaper wrote of him in their leading article:

> J. Carey, the captain, has been a model footballer, technically efficient, thanks to hard work, a fighter to the last without ever forgetting he is a sportsman, a steadier of the young and inexperienced, an inspirer of the old and tiring, and at all times the most modest of men, though he has won every football honour open to him.

No one, in any sphere of sport, could ever hope for greater commendation.

Born in 1919, Johnny went to Westland Row Christian Brothers' School, where he played Gaelic football before moving on to Home Farm, that most famous of Irish soccer nurseries. From there he went to St James's Gate in 1936, where he operated as an inside forward. And it was in that position he received his first international cap against Norway in Dublin in 1937. After playing six times for "The Gate," he was discovered by Louis Rocca, Manchester United's chief scout, and transferred to United in a £200 transaction. Mr Rocca, who was associated with Manchester United for an incredible fifty-five years, said of Johnny:

"No greater Irish player crossed the channel to make a name in English football."

Considering the present day values placed on footballers who haven't got half his ability, he would be a "steal" at £10 million. But the real Johnny Carey legend was born when United transformed him into a full-back. It was at right full-back that he earned every honour the game had to offer, including the rare distinction of being invited to captain the Rest of Europe team against Great Britain in 1947. 1949 provided two highlights for Johnny when he was named Footballer of the Year in Britain and played his part in Ireland's memorable 2 - 0 victory over England in Goodison Park.

Johnny, who played in ten different positions during his career with Manchester United and the Republic of Ireland, was a regular member of the Irish team for sixteen years, during which he gained 29 caps. In one famous weekend, Johnny, Peter Farrell, and Con Martin played for Northern Ireland against England in Belfast on the Saturday, and then for the Republic against England on the Monday. He was capped seven times for Northern Ireland. In all, he played 344 league and cup games over his seventeen years with United, and would undoubtedly have played more had not his war service robbed him of six years' league football. Being from neutral Ireland, he could have come home to the safety of this country when war broke out, but instead he enlisted with the Queen's Royal Hussars, later saying, "A country which gives me my living is worth fighting for."

After finishing as a player, Johnny, or Gentleman John, as he was known, went on to make his mark as a manager with Blackburn Rovers, Everton, Leyton Orient, and Nottingham. But in 1972, at the comparatively young age of 53, he gave up football management and went into local government work in Manchester, from which he retired at age 65. In 1953 Johnny played his last game of football for Ireland, against Austria in Dublin, when, in the position of centre-half, he captained his team to a magnificent 4 - 0 victory. He went on to manage the Irish team for many years after that. One of Johnny's most cherished possessions is his F.A.I. Youth Cup medal, won with

Home Farm. Another honour he greatly appreciated was when Westland Row C.B.S. Past Pupils' Union elected him "Row Man of the Year" in 1987. In 1991, Johnny Carey was elected to the Football Association of Ireland/Opel Hall of Fame, an honour he richly deserved.

Mr Carey, the baker man, was proud of the achievements of his talented footballing son, and so too were all of us in Cork Street.

On the 23rd August 1995, Johnny Carey passed away at the age of seventy-five. He was the consummate professional, talented, even-tempered, an inspiration to fellow players, a sportsman, and a gentleman on and off the field. He gave a lifetime of honest endeavour to a game he loved, a game that transcends all frontiers and is enjoyed the world over, regardless of race, colour or creed — from the poorest barefooted players of the third world countries to the millionaire class.

The pampered, hero-worshipped players of the '90s may walk in his footsteps through Old Trafford and other such temples of the sport, but whether they are good enough to fill his boots — let alone to have laced them up for him — is another matter altogether.

The Horse Trotting

If thoroughbred horse racing is the sport of kings, then horse trotting in my younger days was the sport of the ordinary people. With working horses playing such an important part in our daily lives, everybody took an interest in them. Dubliners in particular flocked in their thousands to the trotting races.

In the 'Forties, my father took me for the first time to the horse trotting races at Raheny. At that time they were organised by the All-Ireland Trotting Association. And there began for me a love affair with the sport which has lasted to this day.

Before trotting began at Raheny, the sport had already enjoyed immense popularity, particularly at the Jones's Road cinder track, which sometimes drew attendances in excess of 30,000. Part of this venue later became Croke Park. On one occasion, back in 1911, Paddy ("The Runner") Fagan from Smithfield ran a foot race against the bay mare *Kathleen H* owned by John Horan, a butcher from Camden Street. Paddy, a first-class athlete who was no stranger to international competition, failed to win, despite a handicap. Other venues such as Shelbourne Park and the Phoenix Park flourished around the turn of the century, such was the popularity of the sport.

My most special memories are of the Raheny races. When my father first took me to Raheny as a little fellow, it was a sleepy little country village, miles from Dublin, which only came awake on Sundays when the colourful trotting people with their magnificent horses came to race. The races were held between spring and autumn each year, and they lasted at Raheny from about 1945 to 1958. My father loved the sport and went to most of the meetings, driving there with his horse and trap, with me, and sometimes the Ma. The excitement of the races would start for me a few days before the meeting, with my Dad and me speculating on likely winners. My Da knew

Jack Murphy, the official starter and announcer for the meeting, and so he was sometimes able to get a programme in advance of the races, which the two of us would pore over for hours. We discussed the handicaps which had been set by Peter Ennis, the official handicapper, and argued as to their fairness, and their likely impact on the finishes, and this of course raised the level of excitement another notch. Mr Ennis was also a master farrier, a title he won in open competition in London, and his Gardiner Lane forge saw many a thoroughbred racing horse being shod there, as well as trotters.

On a race Sunday, the horse, harness and trap, which had been prepared earlier in the week, were finished off, prior to being harnessed together. We had an early dinner, after which we left in plenty of time for the two-hour trip to Raheny. Long before we left, however, another ritual was observed, which added to the excitement. My Dad and I would stand at our gate to see the two trotting horses, *Louie D* and *Laura*, owned by George and Pat Barry of the Greenhills, Tallaght, being walked past our house on their way to the track. I often wondered how those horses, having walked such a long distance to the track, could ever be expected to race when they got there, let alone win.

On our return from the races, having fed and bedded down our own horse, the Da and I would again stand at the gate to see the two horses returning home. It was always a thrill when the drivers pulled up and talked with my Dad about how the races went, giving us some inside information as to what went wrong, what happened in the paddock, who said what — information which the ordinary racegoer would not be privy to. It was great when the Barrys allowed me to hold the reins of the horses, or just pat their necks. The Barry horses raced in the Second-Class Division, and I remember them only because they passed up and down outside our house, and certainly not for their prowess on the track.

The last thing which happened before we set off was that my mother put a big flask of tea, and a pack of sandwiches and home-made cakes into the trap. By the third race, after all the excitement and the sea air, that food parcel had disappeared.

As we neared Raheny village, after the long slow drive from Cork Street, for my Dad would seldom press his horse beyond a comfortable trotting-walking pace, I could hardly bear the excitement. By this time most of the trotting horses would have already arrived at the track, but others were still arriving from every direction. Some were driven along in their racing sulkies, others led by the head, and some others were brought in behind traps and carts. As we met the horses on the road, I would call out to the handlers, asking them what were their chances of winning. When I eventually realised that the answers were nearly always the same — a knowing kind of nod which I took to be a tip which never came up — I wised up and stopped asking.

As we turned left at The Manhattan Bar in Raheny, now only about two hundred yards away from the track, the buzz increased further, as supporters arrived from all directions, loudly cheering from their horse-drawn traps and carts. Others came on bicycles, buses, trains, and walking in groups. We crossed over the hump-back railway bridge at the station, in to a lovely leafy country lane, and there, another fifty yards up on the right-hand side, was the track. As we queued with drivers of traps, carts, milk floats, and other strange-looking wheeled vehicles, there was a great atmosphere, with people calling good-naturedly to each other, passing on tips, calling out the names of their favourite horses and drivers. Once inside the ground, we headed left for our favourite spot, where there was a tree to which the horse could be tied. From here, just past the winning post, we could get a good view of the races, either sitting in the trap, or leaning on the rails.

We always tried to get to the track early, to see the dealers from Dublin erecting their stands and the tents, and the bookies setting up their boards, butter boxes, umbrellas, and money bags. Sometimes, before the horses were called onto the track for the start of a particular race, we would be treated to warm-up runs around the track by some of the first-class racers, and boy it was a thrilling sight. The experts watching the warm-ups would nod and wink knowingly to each other, attaching their own significance to what they had seen, and making notes in their programmes.

As the time drew near for the start of the first race, the late-comers arrived in their droves. The dealers called their wares, the bookies shouted the odds, and the horses, with their colourfully clad drivers, arrived out on the track to warm-up and parade before the race. When the horses had warmed up, Jack Murphy, the starter, would announce:

"Will the horses please get ready to start."

I would nearly wet myself with excitement, and it was many a sandwich that got a quick death at that moment, just to help calm me down. By this stage, all the bets having been placed by the punters, the rails would be jammed with excited onlookers. Then the starter blew his whistle, and commenced the ten-second count. The drivers turned their horses about 120 yards behind their starting marks, and then with a roar and smacking of whips, they thundered down the track, while the starter counted … "six , seven, eight, nine, ten" … they were off and racing.

Those who think that Jack Dempsey and Gene Tunney had some kind of monopoly on "the long count" of ten seconds, would do well to sit through a ten- second starting count at the Raheny races, it was nerve-racking.

The races were run on a "distance handicap" basis, with the fastest horses running off the "scratch" mark, and the less speedy ones starting at different distances before that. A white-coated steward would stand at each handicap starting-point with a flag held down by his side. At the count of "ten" the horse's head had to be behind the mark, otherwise the steward would raise his flag, signalling a false start, and the horses would have to start again. But when everything went right, for both horse and driver, the horse would be absolutely flying just as it reached the starting point.

As the horses sped off around the track, amid shouts from the drivers and cheers from the onlookers, a hail of dark-grey cinder dust would rise up, enveloping all the spectators. By the end of the meeting, we would go home black and in need of a wash. The cinders affected one and all, including personalities like Josef Locke, May Devitt, Noel Purcell, and Hector Grey, who were often seen at the races, cheering their heads off like the rest of us.

What made the races even more exciting for me was that sometimes Dad would give me a shilling to bet on a horse. And if my fancy won the first race, I could bet on the second race, and so on, if I won again. If at the end of the races I showed a profit, my Dad would give me a shilling for myself, thereby ensuring that I would be able to visit the Leinster and Rialto during the week. But even without having a bet, just to pick out a horse and cheer it on for all I was worth was a great thrill.

Many horse owners and supporters were butchers, pig men, cattle men, and dealers of one kind or another. The cattle men were easily picked out by their high-laced tan boots. Any cattle man worth his salt wore these boots, which were hand-made by Barry's of Capel Street. So important a status symbol were these boots for the successful cattle man, that a saying came about: "Tan boots and no breakfast," meaning that they would rather starve than be without their tan boots.

A trotting horse, when in racing gear, is harnessed to a "sulky," a low-slung light-weight, two-wheeled vehicle with a seat on it. The driver sits on the sulky, legs spread-eagled, leaning forward with arms extended holding the reins and a whip. Do not let the term "trotting" confuse you as to the speed at which these horses can travel. Some trotting horses can cover a mile in less than 1 minute 50 seconds. In this country, however, despite ever-improving breeding lines, we have yet to produce a horse capable of getting down to this time. But if the Irish trainers and drivers had the same facilities as, say, the Americans, I feel confident that today's, and tomorrow's stars from the Portmarnock Raceway would be in the top flight of international competition.

Trotting horses trot two ways, either as a "straight trotter," with the right rear leg, and left foreleg movement synchronised, and the same for the left rear leg and right foreleg, which is a natural style, or as a "pacer," where the two left and the two right legs move together, assisted by a harness called "hopples." The pacing style is an unnatural one to the casual observer, but it is as natural as bread and butter if the horse is a pacer, bred from pacing parents. Whereas a new-born thoroughbred racing foal will gallop around naturally soon after birth, a standard pacing foal will pace around. But, for me, of the two styles of trotting, there is nothing more thrilling or exciting as watching a big straight trotter, such as Raheny's *Brugden*, come thundering down the track, head up, with legs stretching high up and out, and the driver just sitting there, arms extended, giving him his head, but not pressing too much for fear of "breaking" him.

Trotting horses are not allowed to gallop, or "break," as it is called. If the horse breaks, the driver has to pull him quickly back to the proper trotting gait or be disqualified. But if the straight trotter can force your heart to miss a beat, through the sheer excitement of the power, speed, and noise of his movement, the one that gets the job done, and wins most of the races on the track, is the pacer.

The pacer, because of its faster leg movements, aided by the hopples, stays in contact with the ground more than the straight trotter, and is unlikely to break, even when put under pressure in the closing stages of a race. The pacer can also negotiate corners more easily that the straight trotter, whose sheer power off the straight into the bends can sometimes force him to run wide, losing valuable ground. In America the straight trotters and pacers usually race separately, such are the large numbers of both styles in training. Here however, the two types race together.

In road racing which covers much longer distances, sometimes up to twelve miles, the straight trotter is more likely to come out on top, as the pacer, bred for racing on a round track, tires more easily, and is at a disadvantage on the straight. There is also a risk of the pacer being cut by the hopples, if raced in for a long period of time.

Before the revival of horse trotting at Raheny in 1945, and indeed for some time after that, a great deal of road racing took place around the Dublin area. Some popular courses were from Ashbourne to Finglas, and Bray to Dublin. Most of the races were against the clock, with horses going off at agreed intervals. There were some "head-to-head" races as well. Because of the long distances involved, these were usually easier races for both driver and animal, as tactics, rather than speed against the clock, became more important. Many of these races came about as a result of open challenges made between trotting men, wherever they assembled. Sometimes owners had to swallow their pride and decline a challenge for they knew it was totally unfair, in that the challenger's horse was much faster than their own, and to race was tantamount to just handing over the challenge money. But mostly these challenges were arranged by middle-men, who genuinely tried to match horses of equal abilities over varying distances, and in a manner which would not cause embarrassment to either owner.

The horse that I remember as being outstanding on the road during the Raheny days was a bay mare named *Grace Dewey*, owned by Barney Ross from the North of Ireland. Indeed the Northern owners played a very significant part in the success of trotting at Raheny, bringing their first-class horses all the way down to Dublin by train to compete at meetings. Some such sportsmen were George, Jim, and Tommy Smith, Crawford Dougall, Sammy Little, James Short, and Barney Ross, to name just a few.

In the Raheny days, the programme consisted of six races, made up of two second-class and two first-class heats, with the first three in each heat progressing to the final. Races were generally run over one and a quarter miles and handicapped on yardage, that is with the fastest horse in each race giving varying distances of start to the other horses. A horse running off the scratch mark in the second-class trot, for example, Tommy Breslin's *Tan Lady*, would concede as much as 440 yards to Danny Norton's *Danny T*. In the first-class trot, Gerry Fogarty's *Emma Lee*, or Joe Plunkett's *The Merry Widow*, both running off scratch, would be asked to concede 150 yards to *No Wonder*, or *Rennie G*. If *Tan Lady*, the second-class scratch runner, had switched to the first-class races, it would have received a handicap of about 160 yards, which illustrates the difference between the two grades of race.

From time to time, races were run over six and eight furlongs. And once I remember a terrific two-mile All-Ireland championship race in which the favourite, *Silver Jubilee*, driven by its owner Paddy Headon, was beaten by the game little pacer *Collector*, driven brilliantly by Jimmy Breslin, who set a scorcher of a pace from the gun. On that occasion, *Silver Jubilee* was raced without the hopples, and this proved the gelding's downfall, for the pressure put on by Paddy Headon to pass *Collector* in the final stages of the race, was just too much for his horse, and it broke, probably for the first time in its life, just as they passed the paddock gate on the last lap, leaving Jimmy Breslin to go home alone over the final two hundred yards.

Nearly all the Raheny drivers were good. The outstanding ones I can remember were George Smith, Jimmy McGuirke,

Jimmy Breslin, Tommy "Boy" Quinn, and Noel Fogarty of *Emma Lee* fame who one day drove the winners of all six races. Ladies also drove from time to time, one occasion being when the international equestrian star Iris Kellett drove *Emma Lee* to victory against an English lady driving *Jimmy Wilde*.

Another great international day was when the magnificent English champion *Hurricane*, a dark brown pacing stallion, and one of the fastest horses ever to have raced at Raheny, came over to take on the best of the home contingent. The Great Northern Railway laid on special trains to take the supporters out from Amiens Street station. If I remember correctly, *Hurricane* ran from 90 yards behind the scratch mark, giving the likes of *Silver Jubilee* a start of more than that distance. The handicap proved just too much for *Hurricane*. Nearly twelve thousand people attended the meeting that day, lining the rails ten deep in places, and if anyone fancied a bet, there were about fifty bookies present to take their wagers. There was another occasion when a visiting circus raced *Circus Pride*, their magnificent chestnut German-bred trotting stallion, which, when not performing under the big top, could fairly fly on the racing track. They were indeed the golden days of modern-day horse trotting in this country.

The horses that I remember as being the fastest from the early Raheny days were Joe Plunkett's brilliant brown mare *The Merry Widow*, Gerry Fogarty's Canadian-bred mare *Emma Lee*, which reputedly ran a mile in two minutes flat in Canada, Paddy Headon's *Silver Jubilee*, Jimmy Breslin's *Collector*, and Joe Plunkett's other horse *Black Prince* and, for my money, the best of all the straight trotters, the mighty *Brugden*, also owned by Gerry Fogarty. Another horse I remember with special affection, is Tim Muldoon's beautiful red chestnut pacer *Gold Flake*. Had there been a prize for the best turned-out horse, driver and sulky, I'm sure that *Gold Flake* would have won it week after week. Then there was *No Wonder*. With a shilling on at 8 to 1, running off a handicap of 150 yards, you were sure to get a run for your money, for off that mark, with Tommy Breslin in the sulky, he took a hell of a lot of catching, and even when caught, passing him was something else. The others I can remember are *Robin Adair*, *Broncho*, *Airbomb*, *TC*, *Tan Lady*, *Louie D*, *Laura*, *Minnie G*, *Big Bill*, and the Captain Wentgesowned *Luxor*, a brilliant horse that came to the track near the end of that era, too late to take on the greats.

Apart from going with my Dad to the trotting, I also got some of my pals interested in the sport. We often cycled out to the Raheny track. For threepence, we could leave our bikes with an attendant who stacked them against a stone wall, just opposite the entrance to the track.

One day, Danny Norton of Bow Lane, from whom we sometimes bought grain for mixing with the pig feeding, gave me a sure-fire tip for his trotting pony *Danny T*. When I passed this information on to my pals, they just laughed their heads off, for *Danny T* at that time was being given a 440 yards start in the second-class races, and still ending up the same distance behind the winner. The thought of *Danny T* ever being able to win a race, no matter what start they gave him, was at that time as far-fetched as the possibility of a man landing on the moon. However, as the week progressed, Danny Norton convinced me that *Danny T* would indeed win on the following Sunday, and at long bookie odds. I, in turn, convinced my pals that this would happen, and we all began to save every penny we could, even staying away from the pictures, to have money to bet on *Danny T*.

Danny T was a humpy little pacer with but two major problems. One was that it had difficulty in getting started in the hopples, and consequently lost a lot of ground at the start, and then galloped more than it paced and in the process usually got itself disqualified. The second problem was that Danny Norton, who drove the little pony, was about 20 stone in weight if he was an ounce, and even with the most finely balanced sulky, pulling the big fellow around the track for one-and-a-quarter miles was one hell of a handicap to impose on any of God's creatures.

On the day of the race we all showed up loaded down with pennies, three-penny pieces, and whatever, to back the certainty. During the pre-race warm-up, we all huddled together on the rails, away from the main crowd, where we could be close to the two Dannys as they came out onto the track. We were hoping that the heavy one would give us the secret nod, to confirm that everything was still in order for his pony to win. The big Danny gave the nod, and a wink to boot. With that signal, we were convinced that a fortune was to hand, and we all hared down to the bookies and piled on our money. No sooner had we placed our bets at 12 to 1, than there was a rush of people "in the know" pouring money on *Danny T*. Suddenly the bookies wiped the odds for *Danny T* off the board, not wanting to have to pay out too much money on a certainty. We knew then that we had made our first killing at gambling, having backed a sure-fire winner at odds of 12 to 1.

As we waited for the race to start, we planned how we would spend our winnings, and the boys thanked me for giving them the tip. But as the horses shot off from their starting marks at the count of ten, the bould *Danny T* could be seen up to his old trick of refusing to get into the hopples. This time though, there was a look of desperation on Danny Norton's face as he tried to get the pony to start. It was obvious to a blind man that the heavier one had put his considerable shirt on the humpy one, and as the rest of the field closed up on them, the considerable shirt, as well as all our money, was about to be lost. Towards the end of the race, the two Dannys began to move somewhat harmoniously, and they scraped into third place from their quarter-mile start, to win a place in the final.

We were broke. Every penny we possessed had been gambled on the Dannys. We were shattered. The new roller skates, the books, the football, the hurley, none of them would now be bought. As the two Dannys trooped slowly by on their way back to the paddock, there were shouts of, "Send him to O'Keeffes the Knackers" and then, "Send them both there." O'Keeffe's was a factory in Newmarket, just off Ardee Street, which processed dead horses into fertiliser, glue and other products.

Shattered from the excitement of the race, or the non-race, depending on your viewpoint, and with not even enough money left to buy a bar of chocolate, we lost interest in the rest of the races and withdrew to lie on the haycocks at the end of the track, soaking up the sun and settling down our poor stricken nerves. After about an hour, *Danny T* came back onto the track for the start of the Second-Class final. A quick trip to the bookies told us that at odds of 12 to 1, there wasn't a penny on him this time.

But at the start of the race, on the count of "ten," little humpy

Danny shot off with such speed, that big Danny nearly fell off the sulky. Round and round the track went the Dannys like a rocket, pulling further and further away from the opposition, eventually passing the winning post before the second horse had even entered the straight. We could hardly believe it. We had backed the humpy one in the wrong race, and although this was no consolation, so had the big Danny.

About 1958, the trotting track at Raheny was closed, and the ground was sold to St Vincent's G.A.A. Club. There was a lot of speculation at the time as to what eventually brought about the demise of the sport. I would say that the plentiful supply of petrol which saw the pig men switching from horse power to motor power, certainly affected the supply of second-class trotters, for many of those horses were every-day working animals when they weren't racing. Whatever the reason, the supporters did not come in large enough numbers to pay the bills. Long before the end at Raheny, owners of the great trotters had also pulled out of the sport.

In 1969, trotting was again revived in Dublin, this time by Hughie Richardson at his Portmarnock Raceway track, and it carries on to this day. Hughie, one of life's gentlemen, built the track slowly and carefully with his family and friends. They laid out a lime-grit half-mile track, similar to those in America, built stables, terracing, stands, a clubhouse, and refreshment rooms. And every Sunday, to this day, from spring to autumn, some really first-class trotting takes place at the track.

Some people say that the great trotters from my boyhood days at Raheny would be no match for the present-day Portmarnock Raceway champions. However, to compare the great Raheny horses with the modern-day Portmarnock stars on times alone would not be fair, nor indeed possible. Whereas the Raheny horses raced mostly over distances of one-and-a-quarter miles, measured from the scratch mark, the Portmarnock horses race a one-mile distance in graded, rather than yardage, handicap races. In addition, the Portmarnock horses have only to race two laps for a mile, whereas the Raheny horses had to run three-and-three-quarter laps on a heavier cinder track for a one-and-a-quarter mile race. As well as that, the fast Raheny horses, coming from the back of the field, running off or near the scratch mark, had to make their way in and around the slower horses that started before them on the yardish handicap system.

Maybe it's just my memory playing tricks on me, or the grey cells disintegrating more quickly than I thought they were, but I just don't believe that any of the modern-day Irish champions could give the likes of *Emma Lee* with Noel Fogarty on board, a start of 100 yards over a mile and win, this being the distance which pro-rata comparison times would suggest the handicap to be. The drivers are easier to compare, Good as the Raheny drivers were, the present-day Portmarnock drivers would appear to be every bit as good.

The trotting at Raheny has given me some very special and loving memories of a sport that was so much a part of Dublin — the horses, the people, the excitement and, most of all, the fun.

If I could relive just one day from the past, it would be a trotting day at Raheny, in the summer, as the sun shone down on the track with Howth Head in the distance, and the haycocks in the top field. I would relive the pleasure, excitement, and fun of getting the black mare and trap ready to go there, sitting beside the track with the mare tied to the big tree, eating oats from the nosebag, as my mother, father, and I picnicked, and the crowd cheered, as *Emma Lee*, *The Merry Widow* and *Brugden* came thundering down the track.

And then we'd go home, with maybe a glass of red lemonade on the way, and a late-night discussion with my Dad about all the excited happenings of the day. I would give almost everything I possess to relive just one such golden day.

THE PRIDE AND THE PAIN

I grew up in The Liberties, but I never knew it was called The Liberties until I moved away from home and out of the area. Perhaps we were an ignorant lot around where we lived, with no sense of history, but that's the way it was, for me. My life in The Liberties was centred around my own dear Cork Street, Weavers' Square and The Tenters where I went to school; Donore Avenue where I went to Mass in St Teresa's; Meath Street and Thomas Street, where I went shopping for the Ma; James's Street, where I occasionally went to the Lyric cinema or passed through on the way to the Phoenix Park; Dolphin's Barn and Rialto where I went to the Leinster and Rialto cinemas; Maryland and the Canal where we went walking; the Greenhills and Tallaght where we played in the sand pits and went on picnics; Francis Street for the Tivoli cinema; and The Coombe, where with an unbelievable innocence, we chased the girls from the Holy Faith Convent when we got big enough to know that they were pleasantly different from boys.

The Liberties has now become a very fashionable place to live. Young professional people, particularly, are buying up old houses, sometimes derelict shells, and spending a fortune refurbishing them, just so that they can have a Liberties address. There was a time when, if you applied for a job with a Liberties address, you stood no chance of getting a reply, let alone an interview. Some of the wrecks, described as "town houses," (a name, I suspect conjured up by estate agents to boost sales) would better serve the community if they were pulled down. Some of the converted properties, sadly, do not recreate or recapture in any way the original design or ambiance.

New luxury apartments and houses are also being built in the area. Some, like the Dublin Corporation's, are tasteful recreations of those pulled down, but others are as much out of place as a pork chop at a Bar Mitzvah. If we wish to retain and

211

renovate what we can of the old Liberties, let us do it to a design in keeping with the traditions of the local architecture. The alternative is to turn a blind eye to the area as we did in the '60s and '70s, when the greedy developers started to pull down the beautiful Georgian buildings.

Particularly in recent times, a lot of books have been written about old Dublin, by writers more knowledgeable on the subject than me. Some have written about the historical and commercial aspects of the area, and others, very simply in their own words, have told what it meant and still means to them. I have written only of part of the city, and the people who lived there, who affected my life growing up — the funny light-hearted things that happened to me, my family, and friends.

But let there be no mistake. As well as the fun, there was also pain. Where I grew up was not a happy place for a lot of its residents, though they were cheerful in their adversity, and as considerate as they could be for the needs of others. Some families lived seven or eight to one room, and existed on bread and tea. Very often it was stale bread discarded by the shops and bakeries, which they would steam in pots to soften. The fires were often cinder fires, the cinders being scavenged from the tipheads where diseases were also picked up. Tea, which was rationed to half-an-ounce per person per week during the war years, was used and re-used until eventually only faintly coloured boiled water resulted.

The killer disease which I remember was tuberculosis, or consumption. If they were lucky, people who developed this were taken away, to some distant hospital. Many did not come back, but the lucky ones did, looking fat and well. Some of those returnees made further trips back to hospital, sometimes not returning the second time.

One of the great fears of Liberties people was that they would die in "The Union," No. 1 James's Street. This hospital (South Dublin Union Hospital) had the stigma of being a poor house, from which people were buried in paupers' graves. The fear of coming to such an end was one of the reasons why the Royal Liver door-to-door insurance agents cleaned up with their penny and tuppence a week policies, which covered burial costs. There was no self-respecting Granny or Granddad who had not provided for a decent coffin from Massey's, or from whoever their favourite undertaker was, as well as a free ride to Mount Jerome and a few drinks afterwards for family and friends.

A real bowsie of a fellow from Maryland died, and out of respect for his decent family, my father and other neighbours went to his funeral. This was a fellow who had caused nothing but trouble since the day he was born. But in the spirit of never speaking badly of the dead, everyone sang his praises at the funeral. My Dad used to say that it would nearly be worth a fellow's while to die, just to hear the good things people would say about him. He also used to say that it does your heart good to see a good funeral.

One of the undertakers in our area who gave good funerals was Massey's of Cork Street. There was a style and solemnity about funerals then, which, unless you've seen one, you couldn't appreciate.

The hearse, which moved almost noiselessly on four large rubber-tyred wheels, to the accompaniment of the clippity-clop of hooves, was either open, or enclosed by glass. A hearse and pair would take the remains to the church on the evening prior to the burial. But the next morning, you would see the real turnout. The hearse was pulled by four horses, and the driver and his assistant, magnificently clad in black livery with silver buttons and top hats, were mounted high on the front of the hearse. The assistant was known as a "mute," just there for show, with instructions to keep his mouth shut. The horses, all black, wore large plumes of feathers atop their heads. If the deceased was married, black plumes were worn and white plumes were used for single people and children. The horses, harnessed in pairs, also wore black or white velvet side-sheets, to match the plumes. The sheets stretched from their withers — the ridge between their shoulder blades along their backs to the top of their tails — being fastened on to the houzen, and the tail end of the crupper. These sheets had variegated designs made from rope, in black or white to match the colours of the sheets and plumes.

The two horses harnessed side by side on the shaft nearest the hearse were known as "wheelers," the ones that really did the pulling, particularly on the flat. The two high-stepping front horses, known as "leaders," were harnessed to a swing bar hooked on to the shaft, and pulled only when going uphill, being mostly for show, just like the "mute." The "wheelers" were sometimes imported from Belgium, it not always being easy to get pure black Irish horses. The Belgian horses were tall, about 17 hands high, with arched necks, and they pranced rather than trotted. They were not considered good for pulling heavy weights, but with so little weight in the hearse, together with their blackness and style, they were ideal for the job. The two leading horses, usually Irish bred, were as black as could be got, and if lighter in colour than the Belgian "wheelers," this was disguised to some extent by the side sheets.

The undertakers also supplied a mourning coach and pair for the chief mourners, as well as a carriage and pair for other members of the family. This was all covered by the basic funeral cost of between £38 and £45, which included the coffin and other extras. Any other carriages and horses supplied would be an additional charge. This price applied in the '40s and into the '50s. The price of a native Irish elm or oak coffin at the time was between £5 and £10.

In 1995 the average price of a basic funeral ranges between £1,800 and £2,500, and it can cost considerably more if a very expensive coffin or casket is required. Most of today's coffins are made from imported wood with exotic sounding names, like African Utile, Obeche, American Oak, and American Mahogany. This is a far cry from the days when I used to run into Massey's carpenter shop with its beautiful smell of wood, and ask if they had any "empty boxes." One day I will get that empty box from Massey's, although I'll be in no condition at the time to go in and ask for it. The one sure thing about this world is that none of us are going to get out of it alive.

When the high-stepping, beautifully turned out horses with the shining hearse and liveried drivers made their way along the street, it was indeed a sight to behold, a worthy send-off for any Dubliner. Shops would close in the area where the deceased used to live, blinds would be drawn in the windows, and people, who might not even have known the deceased, would stop on the streets, bless themselves, and say a prayer, as the cortège passed. After pausing for a little while outside the home of the deceased, the prancing horses held expertly in check by the driver, it was then on to the cemetery.

After the burial, the mourners would make their way to a pub for refreshments, it not yet being the practice for the mourners to go back to the family home for tea and sandwiches. Mickey Joe, the dairy boy, overcome with emotion on the occasion of drinking the deceased's health in an after-funeral booze-up in O'Grady's pub, was heard to say:

"If I could get a funeral as nice as that tomorrow, I'd be off straight away."

One aspect of the undertaker's business which has changed over the years relates to the level of bad debt which exists in the business today. When I was a boy, it was a matter of personal and family pride to ensure that the undertaker got paid, and he did, no matter who else went without. Indeed, it was always felt that the deceased would not rest unless the funeral costs were paid, and might somehow make their way back to wreak vengeance on someone if it wasn't.

Today, bad debts are increasing the burial charges for those honest enough to pay, and it is becoming so expensive to die that people just cannot afford to pop off. Family pride today is not what it used to be in the Dublin of the rare old times.

My family and I were fortunate enough always to have enough food in our bellies, and shoes on our feet. But I saw children go barefoot on the harshest of days, and others with only old rags or sacking tied around their feet to keep out the cold and wet, and protect against injury. I think it was in the '30s that the *Evening Herald*, out of concern for their barefoot newsboys, opened a subscription list to purchase shoes for them. Subscribers had their names published in the newspaper. When the boys were properly shod, any monies left over were directed towards school children. Teachers noticing a particular hardship could apply to the fund for vouchers. These were then given to the children to exchange for shoes in certain shops around the city. Apart from the voluntary subscriptions, funds were also raised with the presentation each year of The *Herald* Boot Fund Concert, with top artists in the country giving their services free. The concert was hailed as "Ireland's Royal Command" show, and artists were anxious to participate. Towards the end of the '70s, with the Society of St Vincent de Paul doing so much for the poor, and increased prosperity ensuring that everyone could afford shoes, there was no further need for the concerts, and so they were discontinued.

What makes me remember The Liberties with pride is the people — the passionate, loving, gutsy people who would share their last slice of bread with you. We all had our pride of course, and were loath to let others know how bad circumstances really were. But we all knew each other's problems, and when help was offered, in the form of a can of soup or a pot of dinner big enough to feed a whole family, the gift would be graciously accepted. Sometimes to put a face on it, to retain some degree of pride, someone might say:

"You shouldn't have bothered, but it is good of you. I've just had my dinner, but it smells lovely, so I'll have another little bit."

There was a sort of game we all played when offering, or being offered a gift. The recipient would say, "No thank you, it's all right, it's all right," about three times, and the donor would say, "Go on, take it, take it," another three times. With this ritual completed, pride would be satisfied, and the gift accepted.

Where you find passionate, loving, sharing, gutsy people, you will also find other less attractive elements of human behaviour — the angry, short-tempered, fighting men and women. I carry with me through life a love of ordinary working-class people, people who, no matter how bad the times are, will help their own. I also carry the memory of great physical cruelty, which the same working-class people inflicted on each other.

I have seen horrific street fights where the blood ran free, and the participants could hardly stand up after the beatings being exchanged, as a crowd of men, women and sometimes even children looked on, cheering. On the corner of Chamber Street and Weavers' Square, I saw a man of about twenty-five fighting his own father, and cutting his face to ribbons, with onlookers cheering him on. Remembering that incident today fills me with the same abhorrence I felt that summer evening all those years ago.

Around Chamber Street and Francis Street, and beyond that in the haunts of the cattlemen and horsy men over in Queen Street and Smithfield, there were known fighting men, who fought for money, or just to build or protect a reputation. The opponents of these fighting Dubliners were often country men up for the markets, who themselves were hard men. My father knew some of these fighters, who, generally, were quiet-spoken and courteous. The fights were almost invariably bloodbaths. But the participants in these arranged fights were sometimes as professional as you would get inside of the ring, In these fights, when one of the contestants was knocked to the ground, there was no counting. The opponent got up in his own time if he wanted to go on with the fight. Sometimes the handlers, out of consideration for their fighter's welfare, would say, "That's it, the fight's over," when it was obvious that their man was outclassed. The truly brutal fights were the non-professional ones born of anger, and engaged in by people who argued over some spur-of-the-moment stupid thing. But in one such fight, I saw two fellows nearly kill each other, and the next day they were walking down the street chatting away to each other, with their heads bandaged, black eyes, and other facial injuries. That speaks volumes for the character of the Dublin people.

But bad and all as the men's bloodbaths were, it was the women's fights which really turned my stomach. They fought viciously, without any rules, although some of them, like the men, made up afterwards. The first time I saw a woman with no knickers on was in a fight outside a pub in The Coombe, just at the end of Meath Street.

It was a Saturday evening in summer, about 8 o'clock, and Attracta, a tough-looking dealer from Thomas Street, had been put out of the pub for shouting her head off, and calling someone name Katty a whore. After about ten minutes of shouting in through the pub door, "Come out Katty, you're nothing but a whore," Katty, an attractive young "bottle-blonde" emerged from the pub. Her appearance equipped her to be a roaring success in the world's oldest profession.

With the fag still in her mouth, she gave Attracta a belt in the face which landed her on her back. A crowd gathered around expectantly.

Katty shouted, "Get up, Attracta, you big fat bitch."

Attracta got up all right, and made a run at the sexy one, while the crowd spilled out of the pub to join the other onlookers in forming a ring. I noticed that bets were being made by some of the cheering onlookers, while the fighters were killing each other. As the two of them rolled on the ground, kicking, scratching, pulling hair, and calling each other names, the Marquess of Queensberry, as far as they were concerned, could have been a potato picker from Donegal.

From where I stood, I had an excellent view of the fight. I discovered that Attracta, the fat one, had stockings just up to her knees. There was a gap of white skin above the knees, and after that a pair of pinkish knickers, big enough for our gang to camp out in, although personally, I would be happier taking my chances sleeping out in the open under a Chernobyl starry sky. As the two women tumbled around, Katty, knickerless, displayed her bare under-carriage She was pinned down by Attracta just in front of me, with her clothes more neck high than waist high. She was held in that position for what seemed like ages, but not quite long enough for those of us innocents who had never seen such a view before. Ever after that, when I hear the phrase "Being in the right place at the right time," I think of Katty. I was fortunate also in that it was a bright summer's evening with excellent visibility.

Katty went on to win the fight, at one stage sitting knickerless on Attracta's face, which is rather ironic when you consider the names Attracta called her. Katty was carried into the pub shoulder high by more willing, wandering hands than there were visible places to grip, while Attracta was put sitting on the pavement propped up against the wall of the pub, panting heavily, her hair and clothes tossed, and a bottle of stout in her hand. This fight was not brutal compared with other women's fights I have seen, but it was a great eye-opener for me, in every sense of the word. I was now a man of the world, no longer needing to sneak looks at the painting of the nude woman hanging behind Miss Gray's hall door, for I had seen the real thing.

A couple of weeks after the fight, I was in Thomas Street shopping for tomatoes for the Ma, and the blonde owner of the vegetable stall turned around to serve me. It was none other than Katty. I stood there, speechless, just staring at her, as she, with a big smile, bent down and said, "Do you want something son?"

I heard myself mumbling, "I saw you fighting."

Then she said, "You saw me fighting Attracta?" Next she called to a woman at the next stall, "Hey, Attracta, he saw us fighting outside the pub." There to my surprise was Attracta.

She called back, "I hope you enjoyed the fight. It'll be my turn to win the next time," and then they both laughed.

I got the tomatoes from Katty, who said kindly, "There's a few extra ones for yourself. Won't you come back and see me again?"

After that I bought all the vegetables from Katty who remembered me each time I called. And although I told some of my pals about the knickerless woman in the fight, I never gave away her identity, for I felt that would have been letting down a friend.

About two years later, Katty suddenly said goodbye to me. She was going to England the next day. For no reason that I can explain, the news of her departure brought tears to my eyes and a lump to my throat as I turned away from her for the last time. Tomatoes never tasted the same after that.

After the Attracta-Katty affair, I never wanted to see women fight again, for they fought too viciously. But I couldn't avoid seeing a funny kind of women's fight in Thomas Street. I had just come out of John's Lane Chapel, after saying a Hail Mary that the hoor who stole my bike would break his neck off it. Two dealers on adjoining stalls were shouting abuse at each other, as a crowd gathered. One suddenly threw a tomato at the other. The recipient of the hot-house vegetable replied with a head of cabbage, which struck the other woman on the head, knocking her off her box. When she got up, she gave your wan a belt of a turnip.

After that they emptied the contents of their stalls at each other, shouting and calling names all the time. When their stalls were emptied, and both of them black and blue from the flying vegetables, they launched themselves at each other, only to be stopped from coming to blows by two men who held them back. Then the woman who had thrown the cabbage started to cry. With that, her opponent of a few moments earlier, put her arms around her to comfort her. Then everyone around picked up the vegetables and returned them to the stalls, although by now the only thing you could use the tomatoes for would be spaghetti bolognese. When I left, the

MEMORIES

My memories are my greatest treasure
With the passing of time they have turned to gold
A family heritage
Enriched through those who have touched me
At times joyously
Other times sadly and painfully
As I move nervously but hopefully along
In the journey through life
Although my memories are distant sometimes dimmed
Increasingly so with the passing of time
I nevertheless pass them on lovingly and honestly
As they were passed to me
To those I love more than words can tell
To laugh over to cry over to wonder about
To cherish to enrich in their own way
Through those who will touch their lives
And in the cycle of life
To be passed on even more golden
To those they love more than words can tell.

MAP OF THE AREA

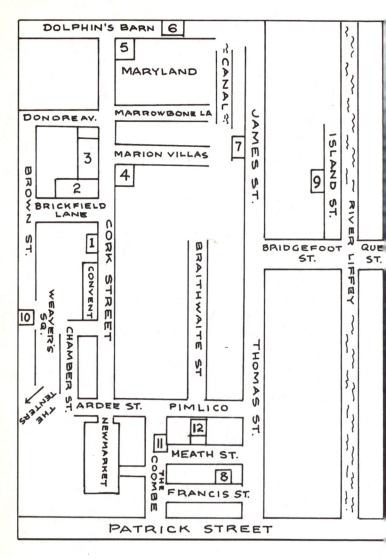

1. Our house and yard
2. Donnelly's factory
3. Fever Hospital
4. Nurses' Home
5. Leinster cinema
6. Rialto cinema
7. Lyric cinema
8. Tivoli cinema
9. The Forge
10. Where Rover fought the cornerboys
11. Where the women fought
12. The Grotto

MAP OF THE FARMYARD

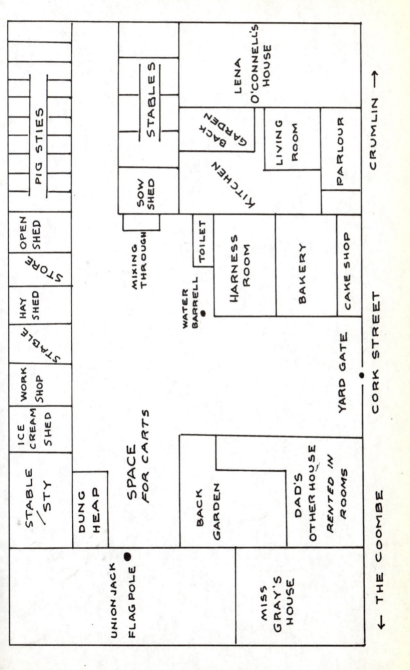